Robot Technology
and Applications

Robot Technology and Applications

Proceedings of the 1st Robotics Europe Conference
Brussels, June 27-28, 1984

Editors:
K. Rathmill, P. MacConaill, S. O'Leary,
and J. Browne

With 80 Figures

Springer-Verlag
Berlin Heidelberg GmbH
1985

Keith RATHMILL
Chairman – Robotics Europe
Cranfield Robotics and Automation Group
Cranfield Institute of Technology
Bedford MK 43 OAL
England

Patricia MACCONAILL
Task Force of Information Technologies and Telecommunications
Commission of the European Communities
200, rue de la Loi
1049 Brussels
Belgium

Patrick O'LEARY
Directorate General III
Commission of the European Communities
200, rue de la Loi
1049 Brussels
Belgium

Jim BROWNE
Secretary – Robotics Europe
Department of Industrial Engineering
University College
Galway
Republic of Ireland

Library of Congress Cataloging in Publication Data
Robotics Europe (Organization). Conference (1st: 1984 : Brussels, Belgium)
Robot technology and applications.
1. Robotics--Congresses. 2. Robots, Industrial--Congresses.
I. Rathmill, K. (Keith), II. Title.
TJ210.3.R63   1984   629.8'92   84-23547

ISBN 978-3-662-02442-3   ISBN 978-3-662-02440-9 (eBook)
DOI 10.1007/978-3-662-02440-9

© Springer-Verlag Berlin Heidelberg 1985
Originally published by Springer-Verlag Berlin Heidelberg New York Tokyo in 1985
Softcover reprint of the hardcover 1st edition 1985

2061/3020-543210

Robotics Europe Conference '84
European Industrial Robot Activities and Perspectives

Conference Organizing Committee

Dr. J. Browne, University College, Galway, Ireland
Mr. J. Chabrol, Renault, France
Dr. R. Dillmann, University of Karlsruhe, FRG
Prof. D. Fabrizi, CEI, Italy
Dr. G. Gini, Politecnico Milano, Italy
Ms. P. A. MacConaill, CEC, Brussels, Belgium
Mr. S. O'Leary, CEC, Brussels, Belgium
Mr. N. Percival, MTIRA, UK
Prof. K. Rathmill, Cranfield Institute of Technology, UK
Prof. U. Rembold, University of Karlsruhe, FRG
Dr. G. Schiele, IPA, Stuttgart, FRG
Dr. P. ten Hagen, Stichting Mathematisch Centrum, Netherlands
Prof. H. J. Warnecke, IPA, Stuttgart, FRG

Editorial Board for the Conference Proceedings

Dr. J. Browne, University College, Galway, Ireland
Mrs. P. A. MacConaill, CEC, Brussels, Belgium
Mr. S. O'Leary, CEC, Brussels, Belgium
Prof. K. Rathmill, Cranfield Institute of Technology, UK

Acknowledgements

The Conference organizing committee gratefully acknowledges the support of the Commission of the European Communities (CEC) for the activities of Robotics Europe over the past two years. In particular, it wishes to acknowledge the support of the CEC for the Robotics Europe Conference '84, of which this book is a record of the papers presented.

Sincere thanks to Cranfield Institute of Technology, in particular to Mrs. Geraldine W. Hookway and her staff, who provided an efficient and courteous administrative and secretarial service for the conference.

Thanks also to Mrs. Geraldine W. Hookway, Mrs. Linda Nicholls and Mr. Joe Van Damme, who provided a liaison and back up service throughout the two days of the conference.

Industrial Robotics Standardization Europe

Industrial Robotics Standardization Europe, known as Robotics Europe is an EEC based organization, supported within the framework of the Information Technologies programme of the European Community.

Founded in June 1982, it has now grown into an active group of some thirty five members drawn from all member countries of the European Community.
The group has now divided into five well defined sub groups working in the following areas:

Health and Safety Aspects of Industrial Robots

Industrial Robot Terminology

Industrial Robot Programming Systems

Industrial Robot Communications and Interfaces

Performance Testing of Industrial Robots.

The objectives of Robotics Europe are as follows:

1. To further the awareness, development and application of industrial robots within the European Community.

2. To pool resources and exchange information on robotics within and outside the EEC in order to ensure that the scarce expertise in robotics and associated techniques is used most effectively.

3. To advise the Commission of the European Communities on robot related topics.

4. To provide an essential link between the EEC Commission and the international and national standards organization within Europe involved in the standardization of robotics and associated technologies.

5. To provide one of the links between the EEC Commission, the United States and Japan for the exchange of information at appropriate level.

This book is a record of proceedings of the first conference of Robotics Europe, held in Brussels on June 27–28, 1984.

For further information on Robotics Europe contact:

Professor Keith Rathmill
Chairman – Robotics Europe
Cranfield Robotics and Automation Group
Cranfield Institute of Technology
Bedford MK43 OAL
England

Telephone: (0234) 750111
Telex: 825072 CITECH G

or

Dr. Jim Browne
Secretary – Robotics Europe
Department of Industrial Engineering
University College
Galway
Republic of Ireland

Telephone: (091) 24411 Ext. 625
Telex: 28823 UCG EI

Table of Contents

X

ESPRIT and Robotics Europe

Patricia MacConaill

Task Force, Information Technologies and Telecommunications

Commission of the European Communities
200, Rue de la Loi - 1049 Brussels - Belgium

The overall goal of ESPRIT is to raise the level of competitivity of European Industry by supporting developments in information technology. Within the domain of ESPRIT, three enabling technologies and two application areas have been identified as being strategically suitable.
These are :

Advanced Micro-electronics
Software Technology
Advanced Information Processing
Office Systems
Computer Integrated Manufacturing (CIM)

In this forum, it is the last mentioned area, CIM, which is of major interest, although clearly there are interrelationships throughout the entire programme.

In CIM, effort is concentrated in two main streams. The first of these is infrastructure, ie., the development of design rules and systems architectures leading to a "reference frame" in which the user can implement his own CIM system in a progressive manner. This topic is considered fundamental to the development of CIM. In general, the existence of a reference model is favourable to users and potential vendors in any particular market. Therefore it is seen as being of particular value to European vendors, who clearly see that any such work cannot proceed in isolation from that of the dominant market developments.

The second stream involves action on those sub-systems, interfaces and tools whose development or refinement is judged to be of strategic value for EC industry (both users and vendors). Exploratory work is planned or underway in the following areas:

- CAD/CAE, - computer aided design/computer aided engineering systems aimed at improving the design process, and at establishing total product models for subsequent use in various stages of the manufacturing process.

- CAM, real-time management of production,

- Machine Control Systems, including flexible machining systems, automated assembly and robotics.
- Sensors and imaging: future systems will need fast analysis of complex images as "sensory input" for CIM applications in areas such as machining, joining and assembly, testing, inspection.

- Micro-electronic sub-systems aimed at integrating machine control sub-systems onto single chips.

Orthogonal to all of these developments will be the
work of integration itself. In this context, areas
where standards, or intercept strategies for standards
are needed will be identified at an early stage. This
in itself would carry obvious benefits for both users
and vendors alike.

In both streams, i.e. infrastructure and sub-
systems/interfaces/tools, developments in other IT
areas, such as micro-electronics and advanced
information processing, will be exploited wherever
appropriate.

Finally, it is considered desirable to establish
centres of expertise, continuing education, and
research, where CIM technologies can be developed and
demonstrated.In general topics are selected where it is
felt there are particular European strengths (eg.
systems design, software development and robot
applications) and hence a good chance of success in
achieving the goals. The value of these topics in
relation to other developments is continually re-
appraised in order to ensure the continued validity of
the programme on a rolling basis.

RELATIONSHIP OF ROBOTICS EUROPE WITH ESPRIT.

There is obvious commonality between the domains of
ESPRIT, CIM and Robotics Europe. The latter is based on
the voluntary efforts of individuals, where meetings
are supported by the Commission, whereas ESPRIT is a
programme of funded projects receiving Community
support up to a level of 50 % of project costs.
For organisational purposes, "Robotics Europe" is run
in close harmony with ESPRIT CIM, and as a result
there is already considerable cross-fertilisation

between the two activities. The voluntary nature of "Robotics Europe" makes it open to a wider participation than might otherwise be the case.

The conference is the first such event organised by Robotics Europe, I would like to thank all those whose efforts have contributed to making it a success.

Performance Characteristics and Performance Testing of Industrial Robots – State of the Art

Prof.Dr.-Ing. H. J. Warnecke, Dipl.-Ing. G. Schiele

Fraunhofer-Institut für Produktionstechnik
und Automatisierung (IPA)
Nobelstr. 12
D-7000 Stuttgart 80
West Germany

Summary

Industrial robots are available on the european market since 1970. But till now only few characteristics are defined to compare different devices. In West Germany the IPA-Stuttgart started in 1974 to develop a test stand for industrial robots. During the last few years many other countries started different research work to develop test facilities for industrial robots. The paper will give you a rough presentation of the defined characteristics and the various existing test facilities in Europe. Additional to the test procedures the paper will show some results which have been worked out at the IPA-Stuttgart during the last years.

Enclosure

In 1975 the IPA-Stuttgart started a research project to develop a test stand for industrial robots. Until now, about 35 to 40 different robots have been tested. Simultaneously, the IPA, together with the VDI (Verein Deutscher Ingenieure) worked out robot performance characteristics as well as measurement guidelines. The following paper will give a survey of this work in Stuttgart and, in a rough summary, of the work of the members universities and manufacturers in other European countries.

1. Testing of Industrial Robots

An industrial robot may be described by approximately 100 pieces of technical data. However, only a part of these characteristics have been yet defined. Fig. 1 gives a survey of characteristics already defined as well as of definitions being worked out presently.

Geometric Parameters	Load Parameters	Kinematic Parameters	Accuracy Parameters
Definition of system limits	Nominal load	Speed parameters	Position accuracy
Definition of spatial subdivision	• Tool load	• Speed	• repeatability
• Danger space	• Net load	• Resultant speed	• positioning accuracy
• Motion space	Maximum net load	• Path speed	• orientation error
• Safety space	Maximum load	Traverse time	Path accuracy
• Non-usable space		Cycle time	• despersion error
Working range			• path deviation
			• circle accuracy
			• corner accuracy
			• rounding off error

Fig. 1: List of Performance Characteristics

The definition of robot characteristics shall enable the comparison of industrial robots of the same or of a similar design. However, it is not only necessary to define these characteristics theoretically, but also to develop procedures to measure these parameters. In order to achieve comparable conditions during the measuring process and, as a consequence, comparable results, the IPA Stuttgart has installed a test stand for industrial robots (see Fig.2) The mechanical part of the test stand consists of a surface plate and a three coordinate measuring device. The measuring device can be moved on three sides of the surface plate on which the industrial robot to be measured is fixed.

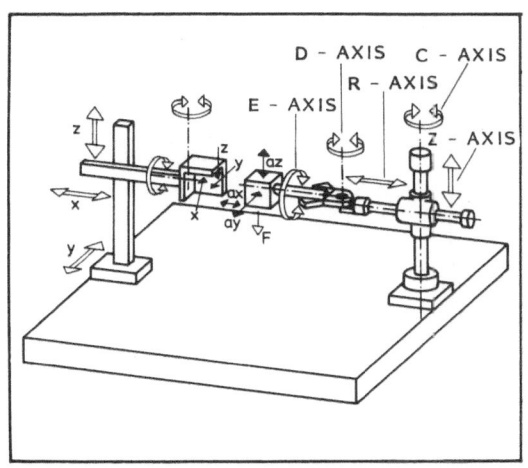

Fig. 2: Test Stand for Industrial Robots

2. Static Measurements

2.1 Definition of Repetition Accuracy

Characteristic measures for quoting the repetition accuracy for positioning and orientating are the mean position dispersion error, the mean orientation dispersion error and the mean reversal error. These parameters are to be stated independently of the basic reference coordinate system and referred to the centre of gravity of the nominal load, Fig.3.

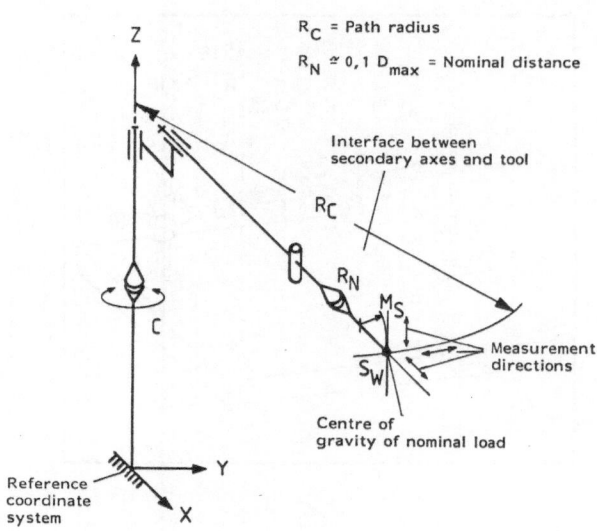

Fig. 3: Geometrical Conditions with the Indication of the
        Repetition Accuracy of Position and Orientation
        (Example: Handling Unit in Cylindrical Coordinate
        Configuration)

2.2 Measurement Principles

One-Cube-Method (IPA-Stuttgart)

At the IPA in Stuttgart a three-dimensional measuring head
with three contactless sensors (position) and another head
with six contacting sensors (orientation) have been developed.
These measuring heads can be fixed at the measuring device as
shown in Fig. 4.

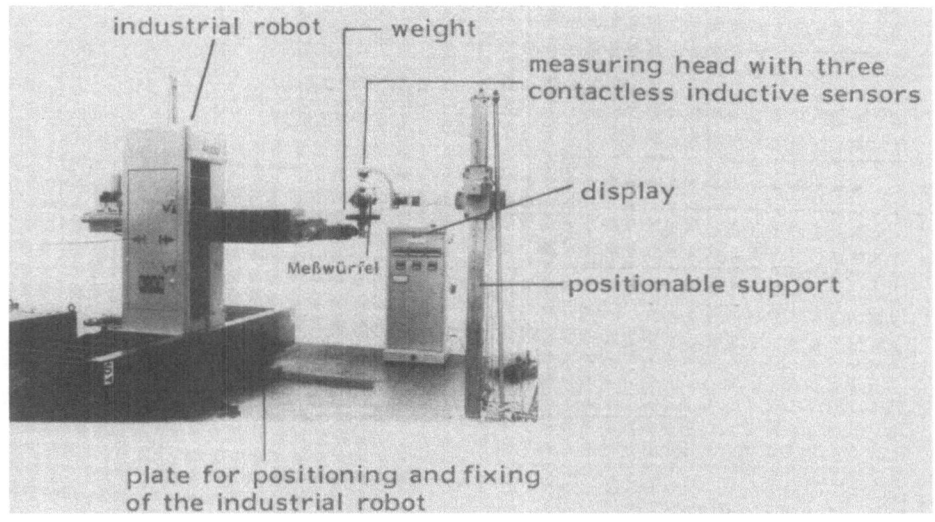

industrial robot — weight

measuring head with three contactless inductive sensors

display

Meßwürfel

positionable support

plate for positioning and fixing of the industrial robot

Fig. 4: Measuring the Position Accuracy with a Special Measuring Head.

The special advantage of this contruction is that it is possible to carry out an absulute measurement without complicated calibration of the whole system. It is also possible to measure the overshoot. The overshoot of an industrial robot is registered during the movement into the final postition. Amplitude and frequency of the distance signal are of interest as well as the time until the amplitude falls below a certain given allowable quantity. (Fig.5.)

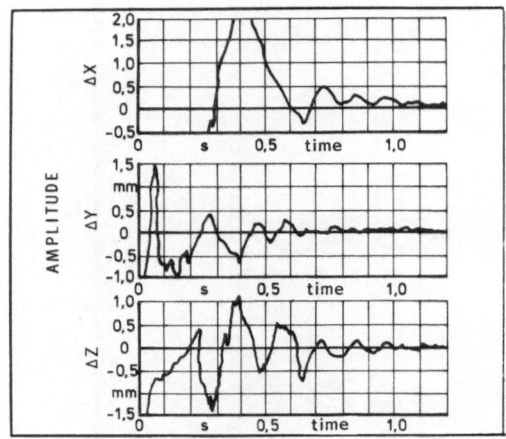

Fig. 5: Overshoot

For measuring the positioning accuracy of an industrial robot, special measuring systems have also been developed in France since the mid 70's.

More-Cube-Method (Citroen-France)

The measuring intrumentation consists of a set of cubes, five for example; each of these cubes, the position and orientation of which are adjustable, is mounted at the end of a rod, the other end of wich is secured onto a rigid plate. All cube-rod sub-assemblies are identical and their orientation with regard to the base plate is adjusted and locked by the operator. The measuring end effector consists of a female trihedron equiped with three sensors on each of its faces. The robot is programmed so that the trihedron is positioned and oriented on the cube faces at the end of each trajectory. (Fig.6) The distances between the opposite faces of a cube and the trihedron are measured by means of nine proximity sensors, and the processing of the data issued from the sensors determines the end-effector position and orientation with regard to the cube

taken as a reference. Repeatability and reversal error tests
carried out according to the above method do not require the
absolute knowledge of each cube position and orientation to be
known in an adequate reference set of coordinates which is attached to the measuring site. This operation requires a preliminary calibration of the measuring site. In a laboratory,
this calibration can be made by means of a 3-D- measuring-ma-
chine. In a workshop, this calibration is carried out through
optical triangulations obtained by means of two theodolites a
multidirectional target precisely secured onto each cube. The
special measuring facilities are outlined in Fig. 6.

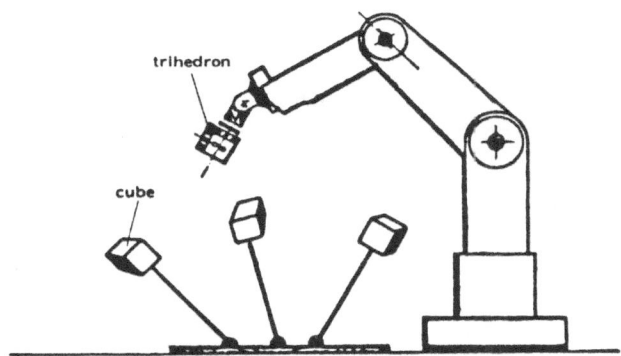

Fig. 6: Measuring the Positioning Accuracy of an Industrial
Robot with the "More-Cube-Method"

Two Theodolite Method

The apparatus (Fig.7.) essentially consists of two theodolites, the respective positions of which are calibrated by means
of landmarks materialized by the graduations of rulers or the
vertices of a calibrated trihedron.

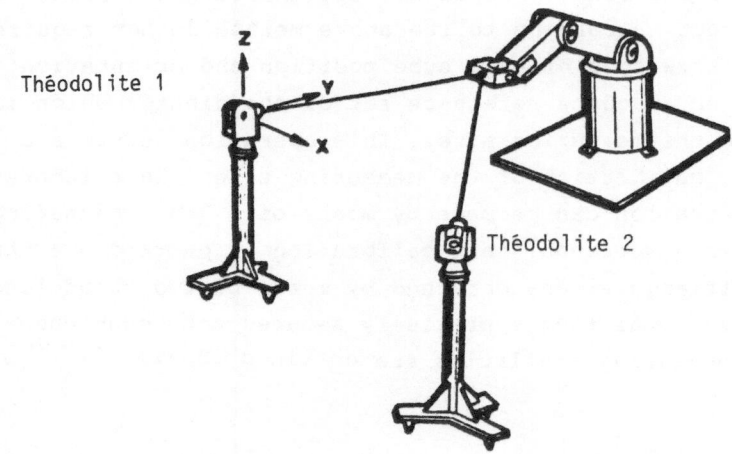

Fig.7: Two-Theodolite-Method

To show the postition of the robot a special light source made
of five arms which are ended by a lightpoint is fixed at the
robot (Fig.8.)

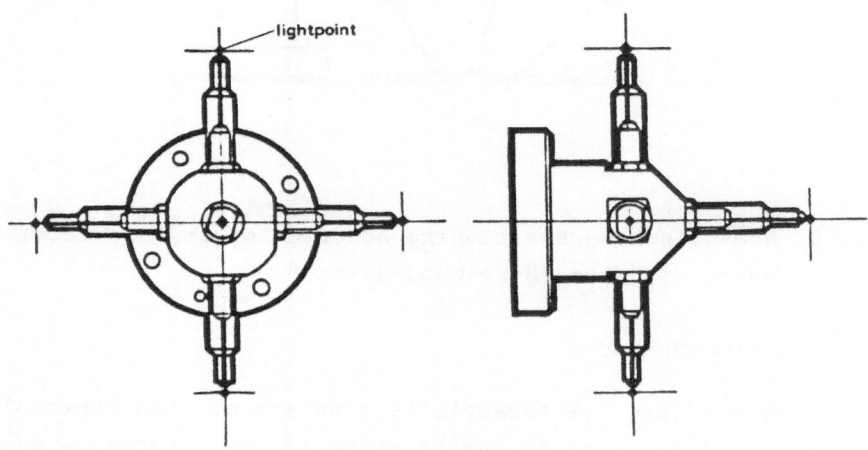

Fig 8: Principle of the Measuring End Effector

To measure the position of the lightpoints in the space of the robot, both theodolites are aimed at the same lightpoint. Then azimut and elevating angles values are recorded. This operation is repeated for three rods of the measuring end effector and the processing of the collected data permits the position and orientation of the end effector to be determined.

3 Dynamic Measurement

Nowadays more and more robots have a continous path control. An analysis of an IR-catalogue is shown in Fig.9.

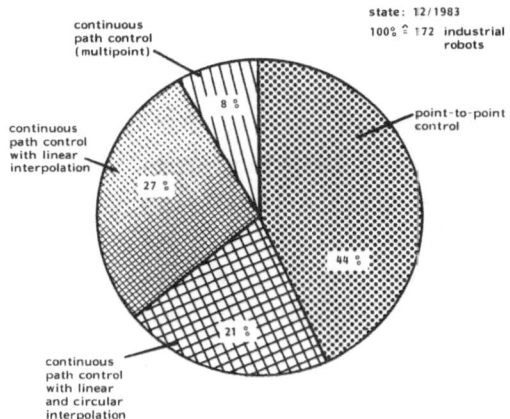

Fig. 9: Control Systems

This demonstrates that the measurement of the dynamic accurary along a straight line or a circle is very important. Here the state of the art is to fix a ruler in the space of the robot as shown in Fig. 10.

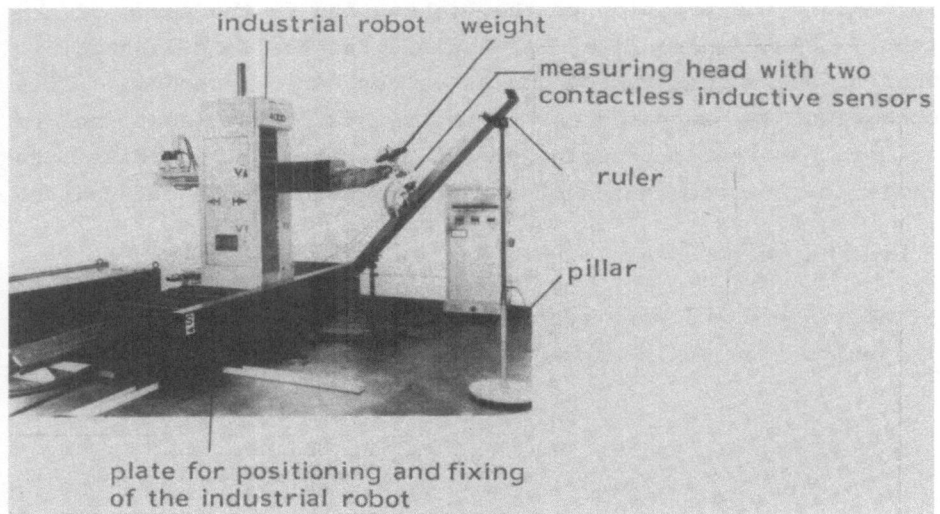

industrial robot   weight

measuring head with two
contactless inductive sensors

ruler

pillar

plate for positioning and fixing
of the industrial robot

Fig. 10: Measurement of a Straight Line

Fig. 11 shows the contactless measuring head (2-D-Type). During programming the sensors are set perpendicular to the surface of the rule, thus registering bucklings which can appear because of movements of the wrist axes as deviation along a required path is registered continuously.

Fig. 11: 2-D Measuring Head

Fig.12 shows the course of a path. Specifications concerning
the velocity along the path, which results from the spread
around an average actual path, as well as about the trailing
error, which is characterized by the distance between the
average actual path and the required path, are obtained from
the graphs.

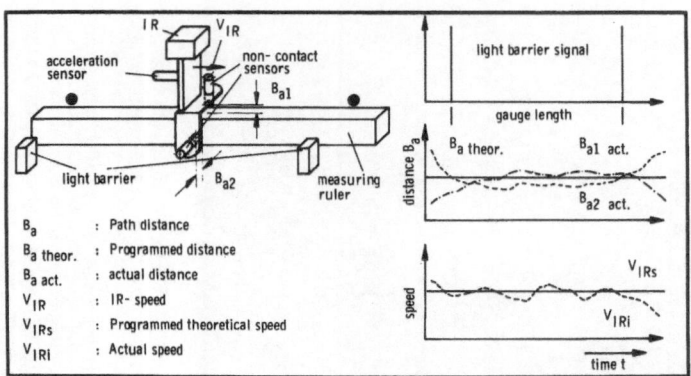

Fig. 12: Result of Measuring the Path Accuracy

Another method to measure the path accuracy is shown in Fig.13.

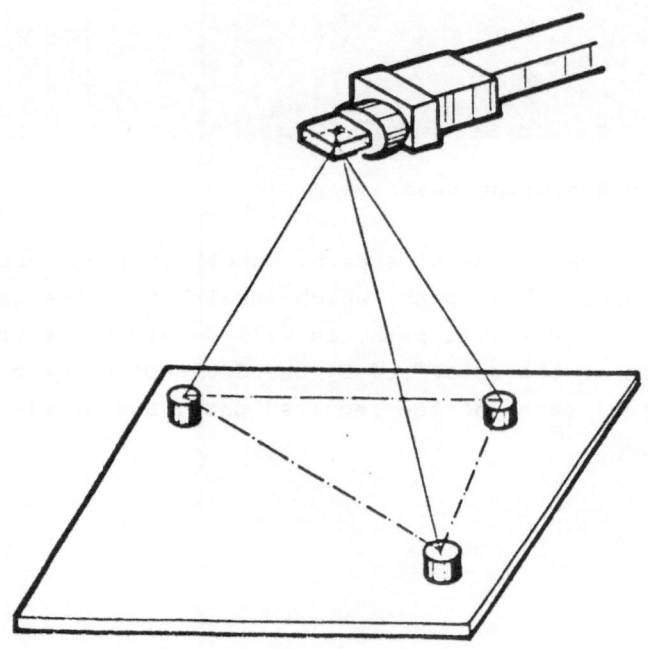

Fig. 13: Three-Linear-Travel-Sensors

This system enables to measure a straight line as well as circle. The measurable length of a path depends on the length of the sensors, the measuring range and the distance of the ball sockets on the plate.

Conclusion

This paper gives a survey of the state of the art in robot performance testing. The experiences made by the measurements carried out at the IPA are transmitted to national (VDI) as well as to international (ISO) organizations. As fast as possible, guidelines for the manufacturers and users of industrial robots will be worked out. The market of industrial robots shall be, in this way, more transparent.

Bibliography

/1/ Brodbeck B.; Schiele G.: Ergebnisse aus den Prüfungen von Industrierobotern auf einem Prüfstand
Technische Rundschau Nr. 3 (1980), Bern.

/2/ Warnecke H. J.; Schraft R. D.: Industrieroboter Katalog 1984
Krausskopf-Verlag, Mainz

/3/ French Proposal on Industrial Robots
Terminology an Performances
Febr. 1982

Project for Development of a Photogrammetric Method for the Evaluation of the Dynamic Performance of Industrial Robots

Marc PRIEL - Bernard Schatz
Laboratoire National d'Essais (Paris) - D.M.I. -

SUMMARY

Created in 1900, the Laboratoire National d'Essais (L.N.E.) has acquired a long and interdisciplinary experience in the field of measurement and testing. Its activity concerns the quality of industrial products and equipment, and, more generally testing related to the activity of industry. Its test reports are applicable in many circumstances: implementation of technical regulation in the field of safety, certification of quality, calibration, expertise.

In 1981, the Bureau National de Metrologie[*] (B.N.M.) in cooperation with users of industrial robots and the robot industry, set up a research program. L.N.E. was charged to study measurement methods in order to evaluate the in situ performances of industrial robots in both dynamic and static mode. The first step during 1982 and 1983 has consisted of developing instrumentation for static measurements. Additional measuring devices permit testing of dynamic accuracy over linear and circular paths.

For measuring dynamic accuracy of robots for applications like paint spraying, welding, deburring, glue application, L.N.E. aims to develop a shop floor method in cooperation with Renault. The aim is to use two high definition T.V. video cameras. The images of three different targets carried by the end effector and of fixed landmarks will be digitalized, recorded and computed in real time, in order to reconstitute the path in space of the end effector. This project needs important efforts in software development and data acquisition with T.V. cameras. This project will use existing work at L.N.E. (infra-red images processing) and Renault.

It was decided by L.N.E. to begin with preliminary phase consisting of an automatic photogrammetry system. This study will solve different difficulties of the T.V. video camera method. By means of a stroboscopic illuminating source, images of targets borne by the measuring end effector

(*) French organisation for metrology

will be recorded on the photographic plates of two photogrammetric cameras. Then the two plates will be digitalized by a C.C.D. camera. Photogrammetric software will restitute the path in space of the end effector.

1. EVALUATION OF INDUSTRIAL ROBOT PERFORMANCE

Industrial robots are becoming a more and more important aspect of applied production engineering technology. The evaluation of their functional characteristics, feasibility and safety is an essential stage in their development.

The Bureau National de Metrologie (B.N.M.) has charged the Laboratoire National d'Essais to undertake a research programme into the methods and necessary instrumentation for testing of performance, in static and dynamic mode of industrial robots. Between 1982 and 1984 L.N.E.'s work in this area has been principally concerned with performance testing in static mode - work on performance testing in dynamic mode will start in 1984.

2. CHARACTERISTICS OF INDUSTRIAL ROBOTS FOR EVALUATION

As a result of frequent contacts with industrialists, users and manufacturers of industrial robots, either direct or through the medium of the enquiry supported by the Association Française de Robotique Industrielle we have assembled a lot of characteristics. Work on standardisation has given rise to drafting of an experimental standard - E61.103: 'Industrial Robots - Performance - definitions'.

Technical studies and standardisation work are in progress with a view to defining test procedures.

Methods of performance testing in static mode:-

- static accuracy
- repeatability of static position
- reversability
- resolution
- load deflection
- stability
- compliance

<u>Methods of performance testing in dynamic mode</u>:-

- dynamic accuracy
- repeatability in dynamic mode
- acceleration
- speed
- stabilisation time
- cycle time
- response time

3. MEASURING METHOD FOR PERFORMANCE TESTING IN STATIC MODE

3.1 Principle of measuring method

The measuring instrumentation consists of a set of cubes, five for example; each of these cubes, the position and orientation of which are adjustable, is mounted at the end of a rod, the other end of which is secured onto a rigid plate.

All cube-rod sub-assemblies are identical and their orientation with regard to the base plate is adjusted and locked by the operator.

The measuring end effector consists of a female trihedron equipped with three sensors on each of its faces. The robot is programmed so that the trihedron is positioned and oriented on the cube faces at the end of each trajectory.

The distances between the opposite faces of a cube and the trihedron are measured by means of nine proximity sensors, and the processing of the data issued from the sensors determines the end-effector position and orientation with regard to the cube taken as a reference.

Repeatability and reversal error tests carried out according to the above method do not require the absolute knowledge of each cube position and orientation. But testing the robot accuracy requires each cube position and orientation to be known in an adequate reference set of coordinates which is attached to the measuring site.

This operation requires a preliminary calibration of the measuring site. In a laboratory, this calibration can be made by means of a 3D measuring machine. In a workshop, this calibration is carried out through optical triangulations obtained by sighting by means of two theodolites a multi-directional target precisely secured onto each cube.

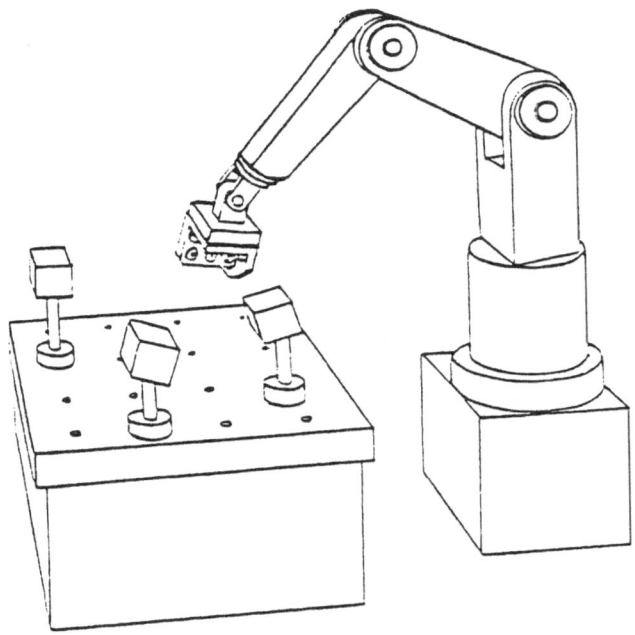

<u>Diagram of a test installation</u>

This required preparatory calibration of the site. In a laboratory this can be done with advantage by means of a tridimensional measuring machine. In a workshop optical triangulations obtained by sightings with two theodolites can be used.

3.2 Measuring end effector

3.2.1 Geometric principles of the operation of the female trihedron

The female trihedron is composed of three light alloy plates assembled to form a 'a corner of a cube'. A mechanism for linkage to the robot is fixed to one of the plates.

On each of the three plates are three proximity sensors capable of measuring displacement without contact. The sensors are protected against shock by a teflon ring.

A system for calibration and setting of the sensors is integrated into the trihedron. This system incorporates six ball joints allowing precise and repeatable positioning of the trihedron in relation to the measuring cube.

With the help of the balls and the three surfaces in an isostatic support it is possible to obtain a virtual orthonomic referential in which the various measurements are expressed.

Each measurement allows determination of equations concerning the three planes and the intersection of these three planes defines the coordinates of a point. The three vectors perpendicular to the three surfaces furnish information on the orientations. It is also possible to calculate the sum of the three perpendicular vectors (orientation of the diagonal of the cube).

The mass of the measuring terminal is 1.450 kg.

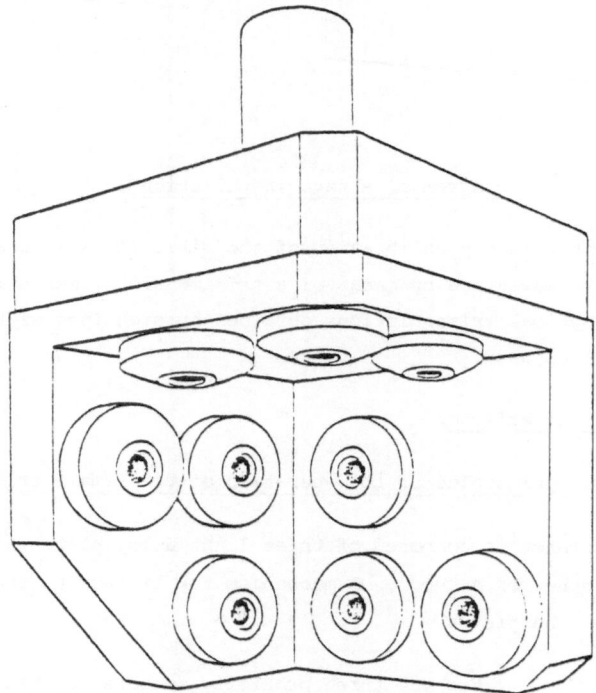

Measuring end effector

3.2.2 Displacement sensors

The choice of sensors was determined by the following characteristics:-

- measuring range 0 to 2 mm (robots to be used for assembly or point to point welding have a varying repeatability of between 0.02 mm and 0.5 mm or even 1 mm)
- resolution of sensors : 1 μm
- passing band : 12 kHz
- measuring force : no contact with the target

We chose to keep to the principle of eddy current sensors as the best for this application.

The main characteristics of the sensors are:-

- measuring range : ± 1 mm
- output voltage : - 4 Va - 20 v
- linearity : 1%
- sensitivity : 8 v/mm
- pass band : 12 kHz (-3d B)

Nine identical measurement channels are installed in the end effector.

3.3 Data collection system

An electronic data collection system consisting of sixteen measuring channels was designed : nine for deplacement sensors and seven for auxiliary measurement for collecting simultaneously other physical parameters (temperature .....).

- Measuring channel of displacement sensors (Nine measurements) The system consists of nine measuring conditions. Nine sample and hold amplifiers
- Measuring channels for auxiliary measurements (seven measurements They consist of seven sample and hold amplifiers with (input ± 10 v)
- Multiplexing and analogue - digital conversion The sampling of the sixteen channels and analogue - digital conversion are achieved by a specific module.

- Principal characteristics of data collection system
  Multiplexing frequency and analogue/digital conversion :
  1000 Hz
- Digitalization on twelve bits
- Transmission by parallel link sixteen bits (four bits
  concerning the address and twelve bits concerning the data)
- The input system is entirely compatible with the other
  sensors without contact having more significant measuring
  ranges

3.4 Data processing and software

The data input system has been coupled to a mini-calculator Goupil 3
(microprocessor 6809). For its own purposes the L.N.E. uses an HP 1000
computer for various different applications (performance testing in
dynamic mode ...).

Currently the following data input software is used:-

- Data input by manual setting or by closing of a contact
  by the robot
- Automatic data input (the software takes account of varying
  parameters - frequency of sampling, stability of robot during
  measuring time)

The software for specific processing has been developed in order to
quantify one of the essential characteristics:- repeatability of
static pose.

The measuring instrumentation allows direct access to the coordinates
(x, y, z) from a point of the measuring terminal. The repetition of a
successive positioning generates a scatter of n coordinate points
(x_i, y_i, z_i).

In order to quantify these dispersed phenomena several methods of
calculation have been used and software developed.

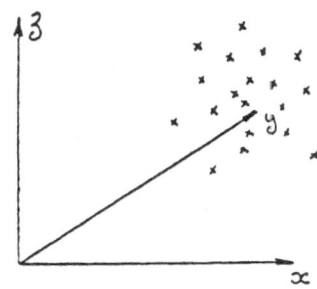

First method of quantification

The parallelepiped containing all the points is studied:-

$$x \min = \inf (xi), \quad x \max = \sup (xi)$$
$$y \min = \inf (yu), \quad y \max = \sup (yi)$$
$$z \min = \inf (zi), \quad z \max = \sup (zi)$$

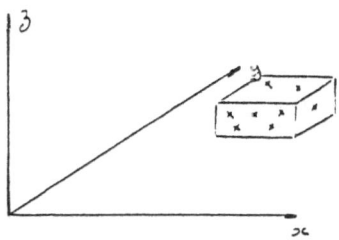

Second method of quantification

The sphere centred on the barycentre of the scatter of points and containing all points is studied:-

- Coordinates of the barycentre:

$$\bar{x} = \frac{\sum\limits_{i=1}^{n} x}{n}, \quad \bar{y} = \frac{\sum\limits_{i=1}^{n} y}{n}, \quad \bar{z} = \frac{\sum\limits_{i=1}^{n} z}{n}$$

- Radius of the sphere:

$$r^2 \max = \max \left\{ (x_i - \bar{x})^2 + (yi - \bar{y})^2 + (zi - \bar{z})^2 \right\}$$

r max represents a value characterising the repeatability

It can equally be calculated:-

$$r_1 = \sqrt{(x_i - x)^2 + (y_i - \bar{y})^2 + (z_i - \bar{z})^2}$$

in examining the low distribution of r

Third method of quantification

The scatter of points can be characterised by its ellipsoid of inertia.

This method of presentation has the advantage of providing an indication of the direction of dispersion of the industrial robot.

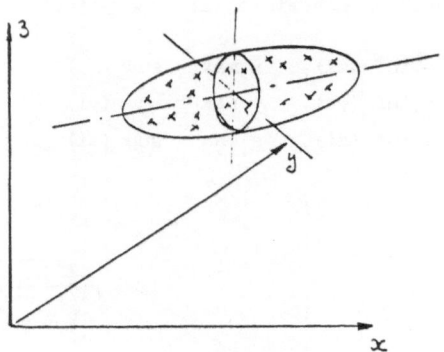

4.   MEASURING METHODS FOR PERFORMANCE TESTING IN DYNAMIC MODE

CHARACTERISTICS	METHODS
Stabilisation time Cycle time Response time	Measuring terminal with cubes for static tests
Dynamic accuracy Dynamic repeatability Acceleration Speed	Linear and circular paths Measuring terminal with designated path
Dynamic accuracy Dynamic repeatability Acceleration Speed	Other paths Photogrammetric strobo-scope

4.1 Measuring method of 3D paths

Numerous industrial robots have systems of linear or circular interpolation and very often complex trajectories result from successive linear or circular movements. We have therefore aimed at extending the scope of the measuring end effector for static testing and follow up for designated paths of this type.

The input system and data processing are identical in principle to those developed for static testing.

A specific measuring end effector is being developed. It will allow simultaneous measuring of distances and orientation in relation to a rule or a disk.

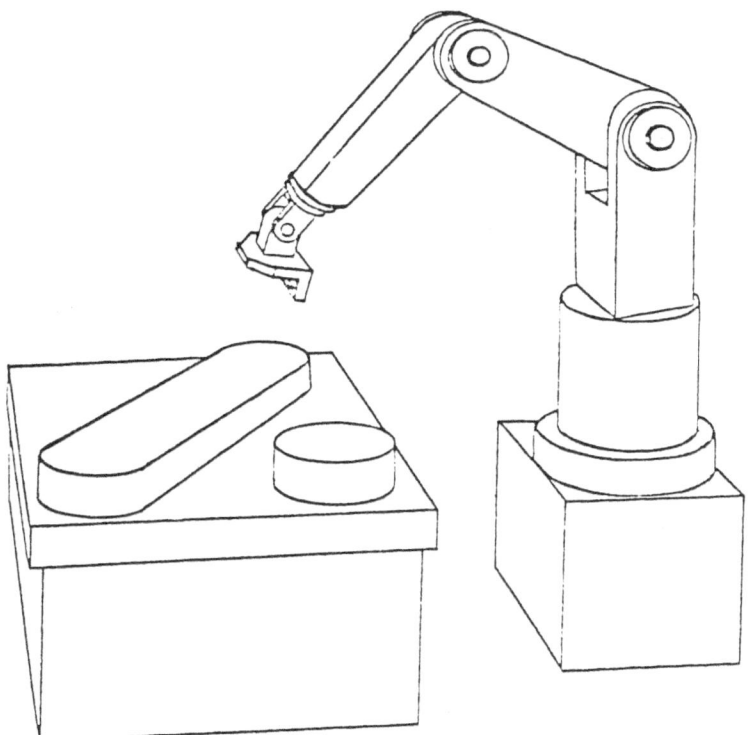

4.2 Methods of measuring other paths

In order to determine the real paths followed by robots including speeds and acceleration the L.N.E. has envisaged development of a method of measurement based on a real time photogrammetric technique using simultaneously two video cameras.

This project poses technological difficulties mainly concerned with camera performance (poor resolution and input speed).

It has thus been decided to put in hand as a first step a photogrammetric stroboscope system.

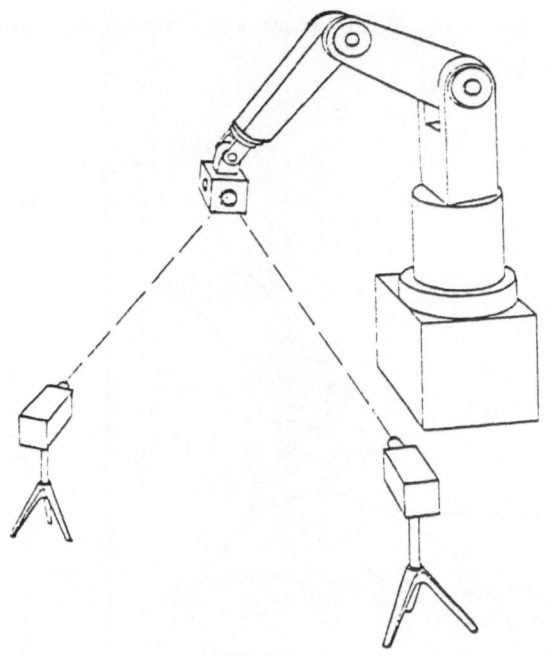

This system will incorporate the following elements:-

- Stroboscope target
  The robot terminal will be equipped with this target which will incorporate a regular lighting system of which the frequency and duration of illumination will be controllable and adjustable according to robot speeds and the degree of accuracy of the measurement sought
- Two photogrammetric photographing cameras
  These two cameras will allow the simultaneous recording of

two photographic negatives in multi-exposure. The number of points recorded along a path will be around sixty

· System of developing photographs

This system will comprise traditional equipment for development of negatives (developer, stop, fixer). Special precautions will be taken to assure dimensional stability of negatives

· Digitalization system for photographic negatives

This system will consist essentially of a camera diode bar allowing an analysis of pairs of pictures. The resolution will be of the order of 2,000 - 3,000 pixels. Coupled to a computer this system will provide coordinates x, y and x^1, y^1 of the equivalent points on the negatives

· Computer

The computer will carry out essentially two tasks:-

- control of the system of digitalization photographic negatives

- data processing and restitution of the path followed by the robot

The following table sets out a test procedure in diagramatic form:-

OPERATION	DURATION
· setting up	30 Mins.
· recording of two photographic negatives	1 Min.
· development of negatives	15 Mins.
· digitalization of negative	10 Mins.
· calculation	
- orientation of various referential (measurements chambers)	10 Mins.
- photogrammetric process	
- restitution of path	

5. CALIBRATION OF VARIOUS MEASURING METHODS AND TRACEABILITY TO STANDARDS

The Laboratoire National d'Essais in its capacity as a calibration centre has various means of measuring, including a tridimensional measuring

machine. This is periodically calibrated and checked. It has been
used for calibration of measuring end effectors. The photogrammetric
photographic cameras will be calibrated with the help of the machine
in setting a net of points.

6.   CONCLUSION

Together these methods will constitute a tool for performance testing
of industrial robots. They are available in the laboratory or on site.
They can be used for type testing and for workshop tests. They will be
available for users and for manufacturers of industrial robots.

Dynamic Performance Measurement of Robot Arms

J. H. Gilby, B.A., R. Mayer, B.Ing. and
G. A. Parker, B.Sc., Ph.D, C.Eng., F.I.Mech.E.
Department of Mechanical Engineering, University of Surrey,
Guildford, Surrey, England.

Abstract.

The paper describes a measurement instrument which is being developed for the dynamic performance assessment of robot arms. A prototype system which can record two degrees of freedom of a rapidly moving target has been constructed and is currently being evaluated. The factors affecting its accuracy and the techniques being used for its calibration are discussed, together with its development into a full three-dimensional measurement system.

Introduction.

Over the last few years there has been a rapid development of a wide range of new industrial robots to meet an extensive range of applications. As familiarity with these devices has grown so have the demands placed upon them and this has given rise to the problem of assessing the suitability of a robot for a given situation. Owing to the geometric complexity of robots the resolution, repeatability, and accuracy may vary within the working volume of the machine even under low-speed operating conditions and the problems are compounded for fast-moving robots carrying large loads. Frequently the information available on these aspects of the robot's characteristics is inadequate and difficult to compare with that provided for other robots. This is particularly so for their dynamic characteristics which are difficult to describe fully by simple parameters.

Ideally, a measurement system for the dynamic assessment of industrial robots should track any trajectory of the arm at full speed with the required positional accuracy thus requiring a resolution which is significantly better than the arm itself.

This implies measuring a trajectory to an accuracy of at least
0.1mm in a 1m cube working volume at speeds up to 5m/s.

The work described in this paper is part of a feasibility
study of a system for measuring robot arm dynamics based on a
laser-tracking technique[1]. This has the potential for fast
tracking together with high positional accuracy, and the
instrument can be configured to operate with different sizes of
working volume.

The Measurement System.

For an instrument to measure the position of a point on a
robot arm which is moving freely in space, it must provide
sufficient information to give at least three equations which
relate the point's co-ordinates. However, additional
information which gives up to three more equations may be
necessary if the original constraints upon the arm position are
not independent of its orientation. Because a measurement
system using a triangulation method can only determine two
degrees of freedom of the target from each measuring location,
at least two locations are required or the method must be used
in conjunction with a direct distance measurement technique
such as interferometry or laser time-of-flight. The
measurement device described in this paper is based on the
first of these approaches.

The scheme of the measurement device with which the
current work is concerned is shown in figure 1. The instrument
consists of a moving target which is rigidly fixed to the robot
arm and two similar static measurement units. Each of these
units (referred to below as a sub-system), consists of both
optical and mechanical components as well as the associated
electronics for its control.

The method of measurement involves following the target
with two separate beams of light (one from each sub-system),
and recording their positions and the distances from their
centres to a known point on the target. This gives sufficient

information for the calculation of the point's position
provided that the relative orientation and separation of the
sub-systems are known. If the displacements of the beams from
the target centre were not measured simultaneously with the
beam positions, then the system would only be able to determine
accurately the robot position when these displacements are
zero; this only occurs, in general, when the target is
stationary because perfect tracking of the target is impossible
to achieve.

To effect target tracking, the distances between the
centre of the target and the centre lines of the beams (the
tracking errors) are measured by light-sensitive detectors.
Error signals from these detectors are used to correct the
directions of the light beams so that they point directly at
the optical centre of the target. This can be done because the
nature of the robot-performance problem permits the use of a
sophisticated target. However, the speed of response of the
measurement system must be sufficiently fast for it to be able
to follow the movements of robot arms. This has been achieved
by ensuring that the instrument has few moving parts which are
of low inertia.

The arrangement of all of the optical components within
one sub-system except the laser head is shown in figure 2 and a
photograph of the prototype sub-system which has been
constructed so far is given in figure 3. The 5mW beam which is
generated by the helium-neon laser is first expanded to about
8mm in diameter and then passed through a beam-splitter and on
to the centre of a plane mirror attached to an optical scanner.
This scanner is able to rapidly rotate the mirror through an
angle of approximately 20° about an axis which is perpendicular
to the beam's direction. The beam reflected by this mirror
impinges upon the axis of a second mirror also rotated by an
optical scanner. The axis of this second mirror is parallel to
the initial direction of the laser beam. After reflection by
both mirrors, the beam emerges from the sub-system with a
direction that is determined by the rotations of the mirrors.
By suitably rotating the shafts of the two optical scanners,

the emergent beam can be directed towards the target as it is carried by the robot throughout the measurement instrument's working volume.

The target of the measurement system is an air-path retro-reflector. Such a reflector consists of three mutually perpendicular plane mirrors and has the property that any ray of light which is reflected by all three mirrors (in an arbitrary order) emerges in a direction which is opposite to its initial direction. However, the reflected ray does not follow the path of the incident way but is displaced by twice the distance between the original ray and the optical centre of the retro-reflector (the intersection of the three mirrors). In other words, the centre of the target lies on a line which is mid-way between the paths of the light ray before and after it passes through the target.

Thus, if the beam of light generated by the sub-system is incident upon the target, then it is reflected back into the sub-system where, after it has been reflected by the mirrors attached to the optical scanner, it is parallel to the initial direction of the laser beam but displaced by twice the tracking error. This displacement of the beam's centre is sensed by a light-sensitive detector after splitting of the beam. (The beam-splitter is necessary so as to remove the detector from the path of the out-going beam).

Using the notation given in figure 4, the equation relating the co-ordinates of the retro-reflector's centre (X, Y, and Z) to the rotations of the optical scanners (θ_x and θ_z) and the horizontal and vertical displacements of the beam on the surface of the detector (η and ξ) is

$$[X,Y,Z] = [-S_z, 0, -C_z]g + [S_x S_z, -C_x, S_x C_z]\frac{\eta}{2}$$

$$+ [-C_z, 0, S_z]\frac{\xi}{2} + [C_x S_z, S_x, C_x C_z]\lambda \qquad (1)$$

where S_x, C_x, S_z, and C_z are equal to $\sin2\theta_x$, $\cos2\theta_x$, $\sin2\theta_z$, and $\cos 2\theta_z$ respectively and λ is unknown. The components of this equation may be re-arranged to give

$$\frac{\eta}{4} = \sqrt{\{(g + \sqrt{[x^2 + z^2]})^2 + y^2\}}\,\delta\theta_x + \frac{x\xi}{\sqrt{(x^2 + z^2)}}\delta\theta_z \qquad (2)$$

and

$$\frac{\xi}{4} = \sqrt{(x^2 + z^2)}\,\delta\theta_z \qquad (3)$$

where $\delta\theta_x$ and $\delta\theta_z$ are small changes in the rotation of the scanner mirrors. The co-efficient of $\delta\theta_z$ in equation 2 is approximately equal to the horizontal tracking error whereas the co-efficient of $\delta\theta_x$ is approximately equal to the distance between the sub-system and the target and so is considerably larger. As a result there is negligible coupling between the equations relating the tracking errors to small scanner movements when the target is stationary and this simplifies the design of the measurement instrument's control system.

Factors Affecting Instrument Accuracy.

The accuracy of the complete measurement instrument is dependent upon influences which may be mechanical, electrical, or optical. Of these, the five principal optical factors are : the system geometry, the directional stability of the laser beams, the perfection of the retro-reflectors, and the accuracy of the optical scanners and the error detectors. Other sources of optical inaccuracy are either negligible (for example the flatness of the mirrors), or are only important because of their effect upon one or more of these factors (for example the stability of the laser beam's intensity distribution is significant because of the characteristics of the error detectors).

Non-ideal behaviour of the retro-reflectors and the error detectors degrade the performance of the complete system because both cause an error in the measurement of the tracking errors of the sub-systems. However, as the error detectors are much smaller than the robot's range of movement, the ratio of this error's tolerable range to the maximum size of the tracking error is much greater than the ratio of the error in the measurement of the robot's position to the size of its working volume. This permits the tracking error to be measured relatively imprecisely.

In contrast, the accuracy of the optical scanners must be better than the measurement system's accuracy because the entire movement of the target across the working volume is translated into rotary movement of the scanners. In addition, the beam emerging from the laser head will not have a perfectly constant direction and, as no provision is included within the measurement instrument for the determination of the pointing drift, this will reduce the maximum permissible error of the optical scanners because both factors contribute independently to the indeterminacy in the directions of the laser beams emerging from the sub-systems. However, the beam pointing drift is considerably reduced by expansion of the laser beam using a collimator; nevertheless this improvement is gained at the expense of a reduction in the effective aperture of the retro-reflector.

The importance of the indeterminacy in the laser beam directions increases proportionally as the robot is moved further away from a sub-system; whereas, to a first approximation, the contribution to the inaccuracy of the complete instrument made by the error detector is independent of this stand-off distance. Thus, not only does the accuracy of the robot measurements depend in absolute terms upon the working volume chosen for the instrument, but the relative significance of the factors causing measurement inaccuracies also changes. There is a further variation of accuracy throughout this working volume owing to the range of the stand-off distance within it and the variation of the angle between the beams incident upon the target. In addition, there are system-geometry errors which are due to the imprecise location of the optical components. Of these the most important is the relative position and orientation of the two measurement sub-systems. This is because the stationary optical components of the measurement instrument are divided into two compact sub-systems within which the position and orientation of the components with respect to each other can be known to a high degree of accuracy and will seldom be altered; whereas, the distance between the sub-systems (normally between two and five metres) is likely to be changed between collecting sets of measurement data as the instrument is moved between sites to assess the performance of different robots or the same one from a different aspect.

The electrical influences upon the instrument's precision include drift owing to temperature, noise, and the accuracy of analog-to-digital conversion. Currently, electrical noise is causing the most serious reduction of the instrument's performance and the electronics are being modified to reduce its significance. This is involving substantial modification of the power supplies and rationalisation of the earthing.

The Calibration Method For The System.

As the relative position and orientation of the sub-systems will be determined whilst the instrument is in use, it is desirable that the system is capable of calculating the relationship between its sub-systems using the outputs which are normally employed in the measurement of target position. Such a strategy avoids the necessity of using the instrument in conjunction with another measurement system. However, it is likely that a calibration target will be required because the sub-systems are usually not positioned so that each lies within the working volume of the other (Indeed, if they were so positioned, the volume in which the complete instrument could perform measurements would be severely limited).

It is not practical for the measurement system to assess its own performance so some form of independent check is essential for the design and calibration of the instrument. As only one sub-system has been built, the problem of assessment and calibration reduces to that of knowing the trajectory of the target with respect to the sub-system's reference frame. Once the target position has been found it may be substituted into equation 1, together with the tracking error, to calculate each scanner rotation and this may then be compared with the experimental data from the position transducers of the optical scanners. As the sub-system has its own controlling microcomputer and data-acquisition system with provision for control by another computer, large quantities of data may be readily collected for statistical analysis with the minimum of operator intervention.

Figure 5 shows a photograph of the calibration rig which is used to move the target; it consists of a carriage carrying the target which is moved along a linear bearing. The carriage is driven by a D.C. motor coupled directly to a lead-screw with a large pitch angle. This gives rapid target movement but crude position accuracy. However, a precision linear displacement transducer of the type used on machine tables is fitted so that even when the carriage is moving at speeds of up to 1m/s its position is known to within 0.02mm. The controller for the calibration rig uses both position and velocity feedback; the former is derived from the displacement transducer and the latter is measured by a tachometer on the shaft of the motor. As data from the calibration rig must be acquired simultaneously with the collection of data from the measurement instrument, it forms part of the system. Currently the hardware and software required for this operation is specific to the calibration rig, but it is hoped that, in future, it may be made more general so that it can be used to record various aspects of the robot's state in conjunction with measurement of its position.

Although the calibration rig gives a known displacement of the target in a straight line, the target's initial position and the orientation of the line with respect to the sub-system must be determined. For this, a theodolite is used to measure a series of points on both the sub-system and the calibration rig. Each point is measured from two locations of the theodolite and its position calculated in the usual manner. However, as the points on the sub-system and on the calibration rig are fixed with respect to each other, the distances between them can be measured by normal workshop metrology methods.

Thus the problem of relating the reference frames of the sub-system and the calibration rig is equivalent to finding two transformations which convert the co-ordinates of a set of points referred to a reference frame defined by either the sub-system or the calibration rig to the same set of points measured with respect to the theodolite's reference frame. Clearly it is advantageous to measure more points than is strictly necessary and then to find the transformation which gives the best fit.

If the points measured in the sub-system's reference frame are p_i and q_i are the corresponding points measured by the theodolite, then the sum of the squares of the distances between the points is

$$S = \sum_{i=1}^{n}(p_i - [Rq_i + \underline{d}])^2 \tag{4}$$

where R is a rotation matrix, $\underline{d}$ is a displacement vector, and n is the number of points. By differentiating this equation with respect to the components of $\underline{d}$ and setting the resultant expressions equal to zero, the following equation may be derived:

$$\underline{d} = \frac{1}{n}\sum_{i=1}^{n}(p_i - Rq_i) \tag{5}$$

This may be used to find the translation vector which minimises S for a given value of the rotation matrix. The derivatives of equation 4 with respect to the Euler angles, θ_j (j=1,2,3), are:

$$\frac{\partial S}{\partial \theta_j} = \sum_{i=1}^{n}2(p_i - Rq_i - \underline{d}).\frac{\partial R}{\partial \theta_j}(\bar{q} - q_i) \tag{6}$$

where $\bar{q} = \frac{1}{n}\sum_{i=1}^{n}q_i$ \hfill (7)

It is not possible to set equations 6 to zero, eliminate $\underline{d}$ using equation 5, and to solve for the Euler angles of the rotation matrix directly. However, the rotation matrix which minimises S can be readily determined using a gradient method because a good first approximation for the Euler angles to commence the iteration can be calculated by selecting three of the measured points and equation 6 can be used to give the gradient.

It is intended that a similar technique will be used when relating the positions of the two sub-systems. If a calibration device which either has an array of fixed targets whose positions are known or is able to move a target to a series of known positions is placed in the instrument's working volume, then the system can collect sufficient information with which to calculate the sub-system positions. The data obtained from such an arrangement is a set of points whose co-ordinates are known with respect to one reference frame (defined by the calibration device) and a set of lines on which they lie with respect to a second reference frame (defined by the sub-system).

Conclusions.

The paper presents a laser-tracking system for the assessment of robot-arm dynamic performance, and the various sources of error which limit its accuracy have been discussed. A bench-mounted laboratory prototype is currently being tested and calibrated, and the preliminary results from these tests are beginning to indicate which parts of the system require improvement. In addition, the electronics of the instrument are being re-built to reduce problems of noise, and a portable version which could be used in a workshop environment is being constructed for three-dimensional measurements. It is anticipated that this will be of use both to manufacturers and users of industrial robots as well as being a valuable research tool.

Acknowledgements.

The authors would like to thank the SERC for financial support for this project under the auspices of the Robotics Initiative. Mr Gilby would also wish to acknowledge the support of a SERC Research Studentship and Mr Mayer the support of a studentship from the Fonds F.C.A.C d'Aide et de Soutien à la Rescherche of the Quebec Province.

References.

1. Gilby, J. H., and Parker, G. A., "Robot Arm Position
    Measurement Using Laser Tracking Techniques." 7th Annual
    British Robot Association Conference (May 1984).

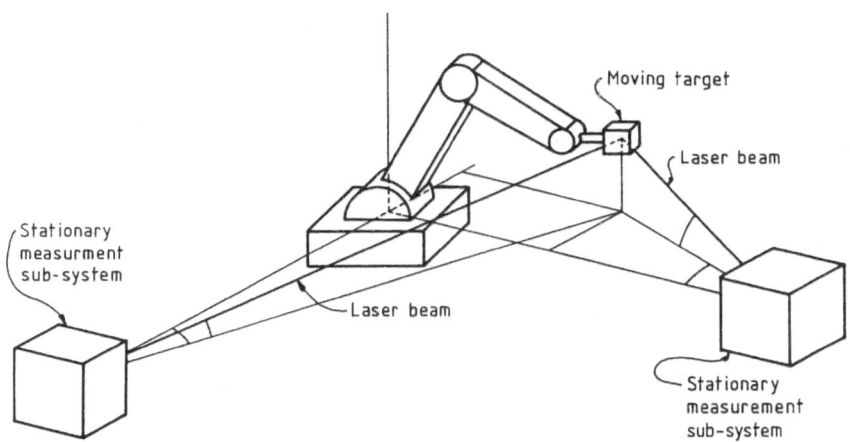

Figure 1.   The Measurement System.

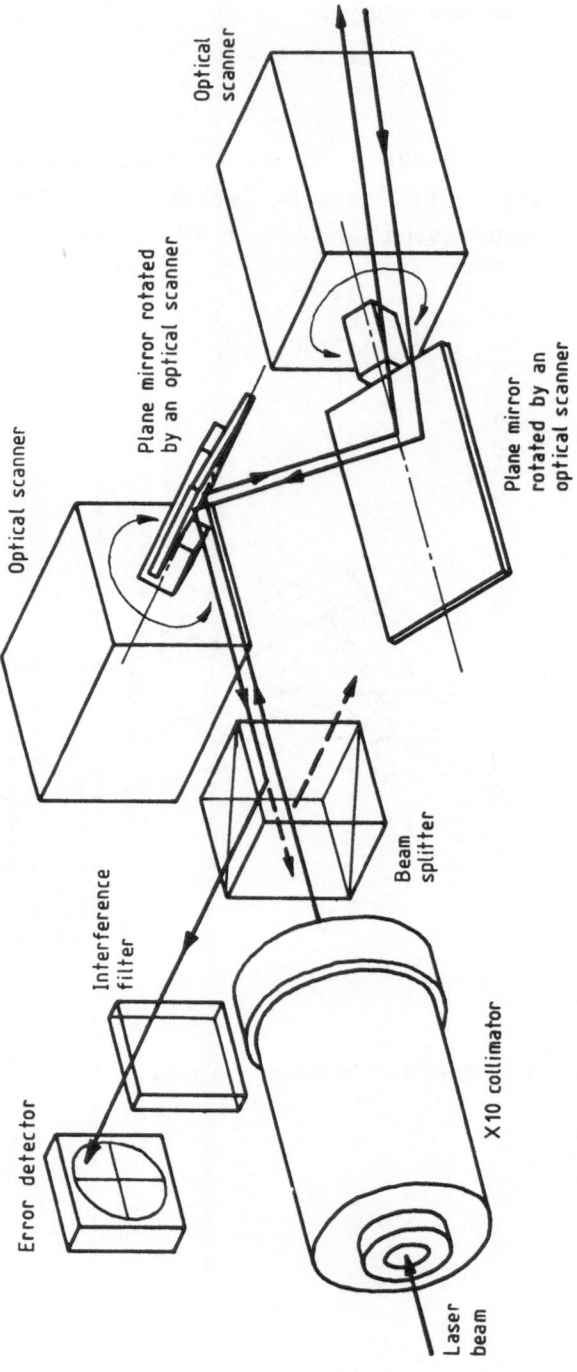

Optical scanner

Optical scanner

Plane mirror rotated
by an optical scanner

Plane mirror
rotated by an
optical scanner

Interference
filter

Beam
splitter

X10 collimator

Error detector

Laser
beam

Figure 2. The arrangement of Optical Components within a Sub-System
of the Measurement Instrument.

Figure 3.   The Prototype Measurement Sub-System.

Figure 5.   The Calibration Rig.

44

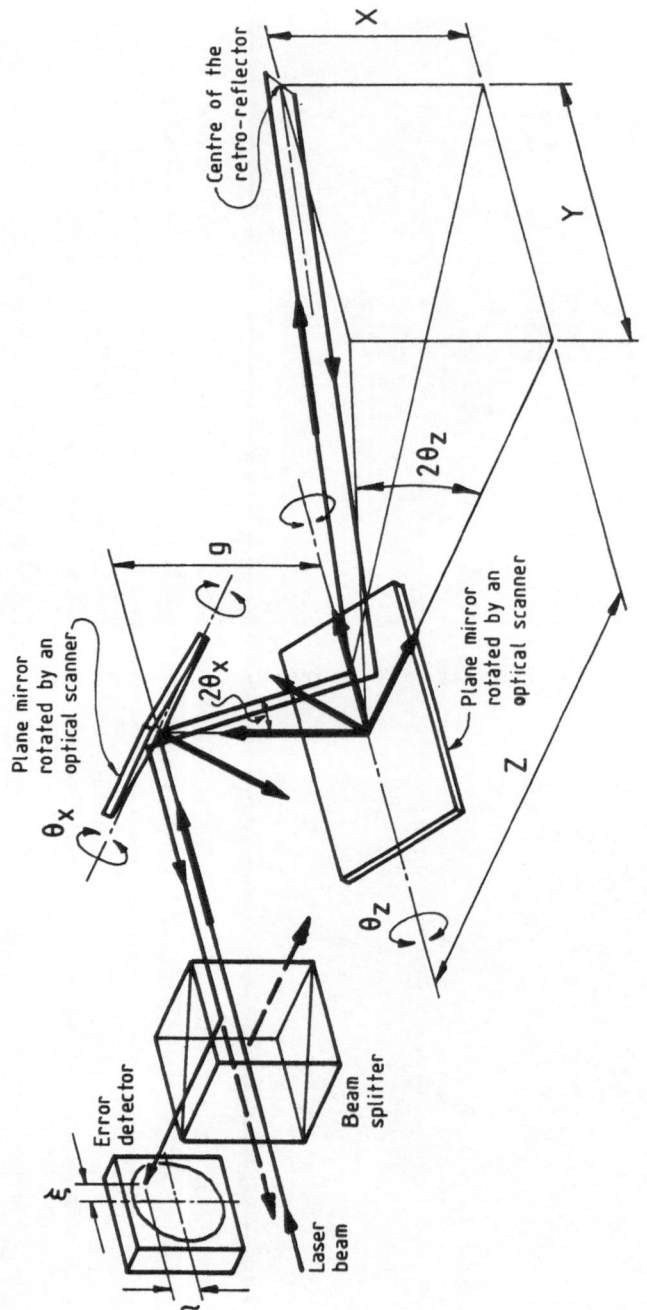

Figure 4. Definition of Notation.

Application of the Experimental Modal-Analysis in the Performance Testing Procedure of Industrial Robots

H.J. Warnecke, R.D. Schraft, M.C. Wanner

Fraunhofer-Institute for Manufacturing Engineering and Automation (IPA), Stuttgart

Summary

Testing procedures for the determination of weak points in a robot structure, overall elastic behaviour and long term performance are discussed. The emphasis in the following paper will be the presentation of the methods and the test results of two different PUMA 560 robots.

1. Introduction

The experimental modal analysis is used to determine the dynamic properties of an industrial robot by identifying its modes of vibration in a specified position and orientation of the robot. In general a mode of vibration is a global property of the structure /1/. A mode of vibration is specified by a natural frequency and a damping factor which can be identified from response data at any point of the structure; furthermore each mode has its characteristic mode shape. Experimental modal analysis is an effort to learn more about the elastic behaviour of industrial robots.

2. Test methods and results

2.1 General

Two Unimation PUMA 560 assembly robots were tested at
the IPA using different test methods /2/, /3/, /4/.
The aim of this IPA-project was the
- comparsion of different test procedures
- comparsion of the same type of robot with a
  different life history. Relation of operational
  hours was approximately 3/1.

2.2 Data aquisition and processing

In this overview aspects characteristic of industrial
robots will be discussed including the most important
steps in data aquisition processing.

a) Test equipment
The requirements for the test equipment are:
- portable in laboratory and shop floor
- all aquired data has to be checked in situ. In most
  cases it is not possible to repeat the testing
  procedure.
- easy calibration for the accelerometers
- simple excitation of the structure
The solution to those requirements can only be a com-
promise similar to the IPA modal-analysis testing
equipment as shown in Fig.1.

Fig. 1: Modal analysis system at the IPA-Stuttgart
including GEN RAD 2505, PCB hammer set and
inductive accelerometers HBM 12/200

b) Excitation of the structure

Of paramount importance in practical application is
the impact excitation with a hammer (force) signal and
the snapback method for heavy structures. The main
advantage of these two methods is the resulting quick
and simple test procedure. It should be noted that
with very low frequencies and nonlinear systems other
methods could be used,   e.g. white noise signal on
one or two of the drives. The reference point of excitation should be carefully checked in view of the
coherence for all important response coordinates.
This is quite difficult for certain robot designs.

c) Reduction in the number of response coordinates.
Before starting extensive data aquisition the number
of measuring points should be evaluated. Experience
has shown that in the first step large numbers are
needed in order to localize specific weak points. In
most cases the arm of the robot is very stiff compared
with the joints so only two measuring points per arm
may be sufficient for the testing in different positions and orientations. It may be possible to reduce
that number depending on the specific design.

d) Data aquisition
After calibration of the measuring system and excitation of the structure the frequency response function
(Bode diagram) will be displayed. Fig.2 shows the
transfer function for a measuring point near the gripper.

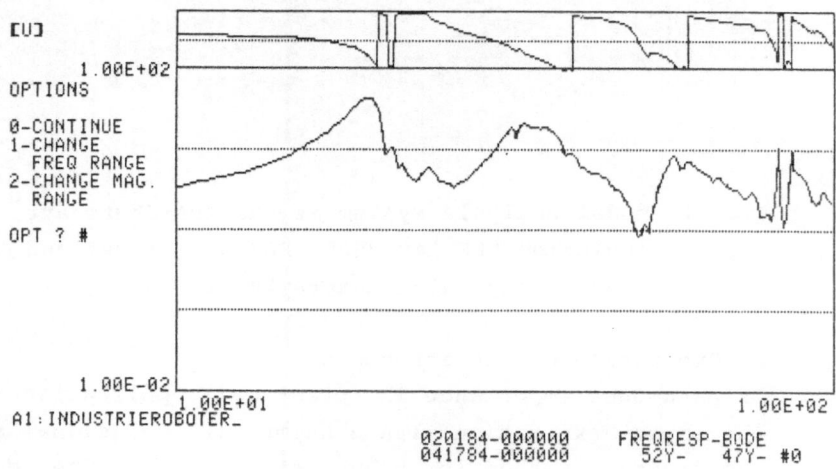

Fig. 2: Frequency response function for a PUMA 560
        robot

Resonance frequencies can be determined using the
peaks in the frequency response function (a/F) and the
shift of the phase. Very important is the coherence
function as shown in Fig.3 for the same measuring
point as in Fig.2.

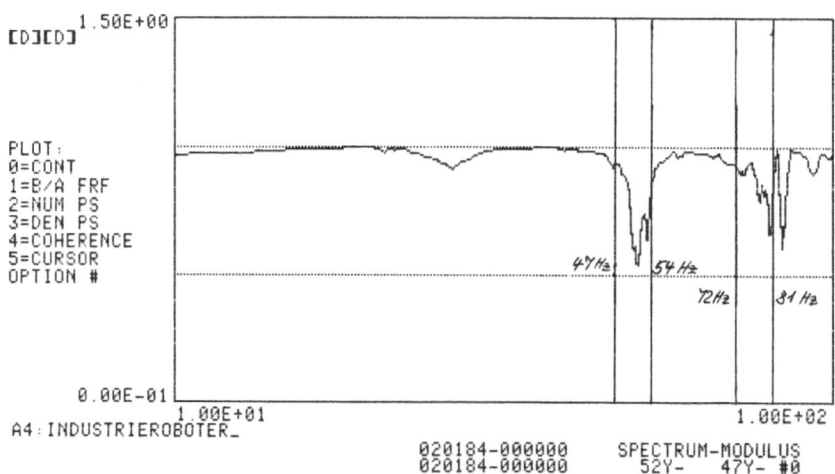

Fig. 3: Coherence function for a PUMA 560 robot

The coherence function is the relation of the response
power caused by the input to the measured response power. The coherence funtion should be one for all frequencies; if not an indication is given about the noise contamination in the transfer function. Other important data for each measuring point could be:
- Numerator and denumerator power spectrum
- Real and imaginary component display
- Phase graph
- Argand and Nichols diagram

e) Modal parameter calculation
An estimation of frequency and damping using the relevant portion of the frequency response function will
be generated and redisplayed using the original frequency
and the system estimation on the same plot as shown in
Fig.4 for the PUMA 560.

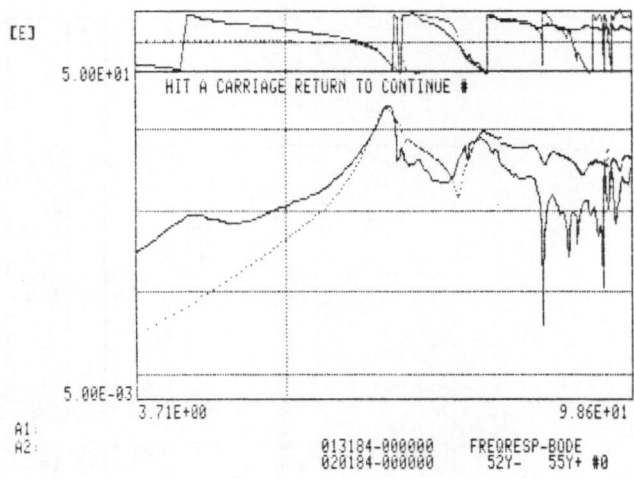

Fig. 4: Frequency response fuction: original and
        system estimation for a PUMA 560 robot

In most cases one particular frequency does not provide all
characteristics of the structure. In this case several frequencies need to be fitted in to get all resonance
frequencies.

f) Mode shape calculation
For the mode shape calculation two methods are common:
- Single degree of freedom technique (SDOF) treating
  each identified mode separately.
- Multiple degree of freedom technique (MDOF)
  treating many modes simultaniously

For industrial robots we start with MDOF in order to
get a clear seperation of the mode which is difficult
in many applications. Then we continue with SDOF for
each measuring point. Fig.5 shows a near perfect
circle fit for the first mode (measuring point near
the gripper)

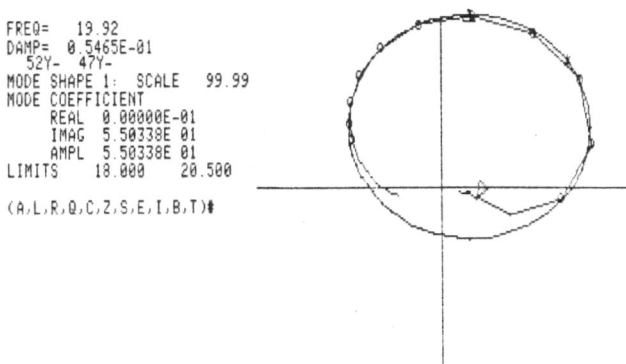

```
FREQ=   19.92
DAMP=  0.5465E-01
   52Y-  47Y-
MODE SHAPE 1:  SCALE   99.99
MODE COEFFICIENT
     REAL  0.00000E-01
     IMAG  5.50338E 01
     AMPL  5.50338E 01
LIMITS   18.000   20.500

(A,L,R,Q,C,Z,S,E,I,B,T)#
```

Fig. 5: Curve fit for the first resonance frequency

Very high damping or a second mode closely coupled for example can be identified by the curve fit technique.

g) Data display
The most important calculated modal parameters will be presented
- resonance frequency
- damping ratio
- amplitude
An example for the PUMA 560 is shown in Fig.6

```
MODE PARAMETERS
LABEL   FREQ    DAMPING  AMPLITUDE   PHASE    REF     RES    MODE    FLAGS
  1    19.918  0 05465    173.8      0.753    52Y-   55Y+    1     0 0 0 1 1
  2    20 999  0.00030    5.986     -2.353    52Y-   55Y+    2     0 0 0 1 1
  3    32.586  0.01796    9.557      1.715    52Y-   55Y+    3     0 0 0 1 1
  4    37.251  0.04658    47.85      0.022    52Y-   55Y+    4     0 0 0 1 1
  5    82.359  0.00173    2.116      1.692    52Y-   55Y+    5     0 0 0 1 1
  6    86.851  0.00792    18.31     -0 260    52Y-   55Y+    6     0 0 0 1 1

HIT A CARRIAGE RETURN TO CONTINUE JQ400:?#
```

Fig. 6: Modal parameters for a PUMA 560 robot

Of considerable interest is the graphic display of the modeshape of the robot in order to find weak points in the structural design. The modeshape can be defined as a vector representing the motion relationship of all points in a robot at a resonance frequency. Fig.7

shows Mode 1 and Mode 4 of the PUMA 560. In Mode 1
the wrist axes and the link between joint 1 and joint
2 can be regarded as weak points, also in view of the
high amplitude. In Mode 4 the foundation and the link
between joint 2 and 3 are weak points.

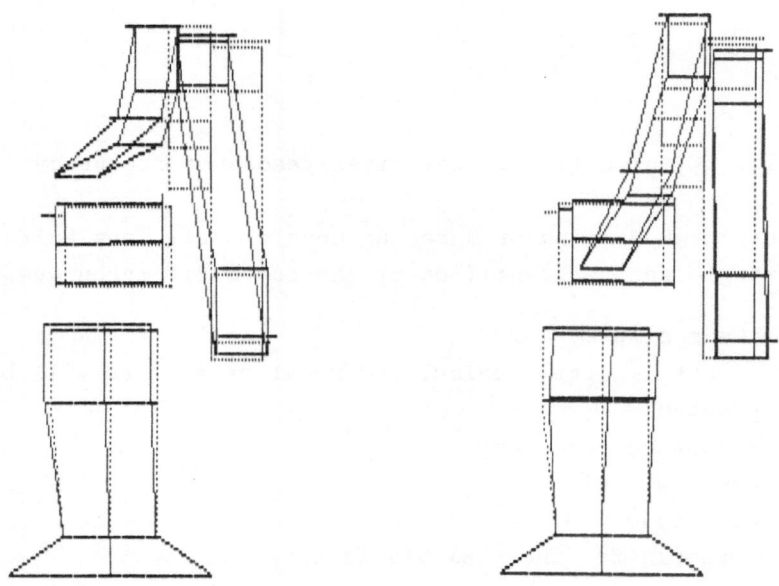

Fig. 7: Modeshape 1 and 4 of a PUMA 560 robot

2.3 Comparison of the same type of robot

Two PUMA 560 with a different life history were tested
under identical conditions. Fig.8 should be compared
with Fig.2 showing the FRF for the same measuring
point. In case of Fig.8 the first resonance frequency
shifted from a close 20 Hz to nearly 15 Hz-similar
phanonena were identified with other frequencies.
This shift was caused by the bearing of joint 2/3 for
the Fig.8 example.

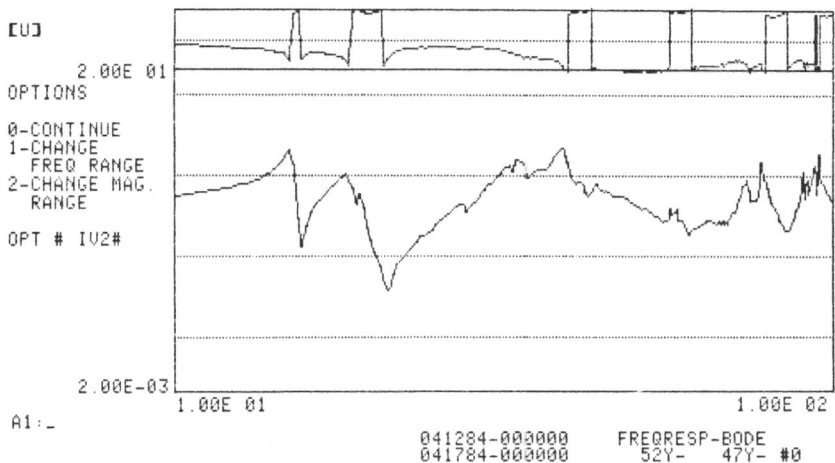

Fig. 8 : Frequency response function for a different
PUMA 560 with a known specific weak point

3. Conclusion

The experimental modal analysis should be regarded as
a powerful tool in analyzing the elastic structure of
a robot during the development process and in the shop
floor. Simplification of the measuring process and
the prediction of mechanical failures are specific approaches for the individual design.

References

1. Richardson, M.: Modal analysis using digital test systems. Seminar on understanding digital control and analyis in vibration test systems (1975), pp. 43-64

2. SDRC DATM/MODAL user manual (1982)

3. Mogler, K.: Experimentelle Modalanalyse an einem Montageroboter. Diplomarbeit am IPA-Stuttgart (1984).

4. Eberle, F., Kadlec, J., Pietsch, C., Scharnowell, R.: Experimentelle Modalanalyse des Industrieroboters PUMA 560, Kernforschungszentrum Karlsruhe TT-Primaerbericht 50.000.0p06A (1983)

A Review of Safety Standards

N. PERCIVAL

Machine Tool Industry Research Association,
Macclesfield, England.

Summary

This paper reviews current developments in the preparation of
standards for the safety of industrial robots.

In the last three years many of the major manufacturing
countries have published standards and recommendations or are
currently actively doing so. This review shows that the
approaches to the subject by the individual nations are very
similar. International work has commenced and the cooperation
of the participants suggests that an international standard
should be available by 1986.

Introduction

It might be thought that with the increasing use of automation
and industrial robots and that having thereby removed the man
from the machine there should be no need to consider the safety
aspects. However robots are different from most other types of
automated plant in that they have unpredictable action patterns.
They are able to move in free space within a working envelope
which is not obvious to the onlooker so that he cannot foresee
what is going to happen next. Also, despite the degree of
automation there is still a need for human intervention for
programming, setting-up and maintenance.

Hazards

Although robots free persons from some of the more traditional
hazards associated with machinery they can present special
dangers. For example the patterns of movement, often carried
out at high speeds, can increase the risk of persons being
struck by moving parts of the robot itself or by parts or tools
carried or manipulated by the robot. Trapping points can be

present on the robot itself and can also be produced by movements of the robot in relation to associated equipment such as
work carriages or transfer mechanisms or by guarding which may
be too close to the working envelope. Other hazards from, say,
electrical shock, burns, radiation etc. can be present depending
on the particular robot application.

Accidents can occur from control errors (either intrinsic
faults within the control system, errors in the software,
electrical interference etc.), mechanical failure (e.g. overloading), electrical or hydraulic failure or from the effect of
the environment (e.g. dust, fume, radiation). Additionally
accidents can be caused by human error by persons working too
close to the robot during programming or maintenance, by
unauthorised access within a robot enclosure or by lack of
familiarity with the equipment.

National standards

Fortunately fatalities have been few although instances have
been reported of near misses and aberrant robot behaviour.
What has been recognised however is the potential for injury
and the need to restrict free access to robot installations.
Many robot user countries have now established rules and
standards for safeguarding robots and although there may be
some criticism of overguarding in some cases it can be argued
that it is better to be safe than sorry.

The initial work on preparing recommendations on the safeguarding
of industrial robots was started in Germany in 1979 with an early
VDI draft and in the United Kingdom in 1980. The latter was
issued as guidance by the Machine Tool Trades Association in
1982 and was quickly followed by East German (similar to the
VDI draft) and Russian standards. Japan issued a standard in
1983 following two fatalities and a government survey of robot
accidents. Recently a revised and considerably enlarged draft
of the VDI recommendation has been published and drafts by
Afnor in France and by the Robot Institute of America are in
preparation.

All of these standards are recommendations rather than being

obligatory. However in Russia and Japan they have led to
amendments in the legislation in these countries. In particular
in Japan the law specifically now asks for measures to be taken
to prevent hazards when robots are used and to educate robot
operatives.

Considering that all the national standards have been produced
independently it is surprising how similar the approaches have
been in the different countries. This may be due not only to
the early recognition of the hazards associated with
industrial robots but also to the cooperation between the
countries involved and their willingness to exchange information
in the draft stages.

All of the national standards give recommendations on the
design, construction and safeguarding of robots. Additionally
the German (FR), American and British documents describe
requirements for the safe installation and use of robots and
give advice on the need for training and education of all
personnel involved. .

Design requirements

The design requirements include the elimination or enclosure of
potentially dangerous parts, use of anti-overrun devices, layout
of controls, power requirements including protection against
irregularities in the power supply and electrical interference
and the need for the robot manufacturer to provide information.

It is the control system design which is currently providing
most controversy from a safety point of view. Most robots are
programmably controlled and if required safety can be built
into the software of the program. Whilst all the standards
ask for an emergency stop so far it is only the British
guidance which specifies that this should be hardwired (although
the German document implies this by referring to other standards
thus providing a higher level of safety integrity). This is
particularly important on teach pendants which not only should
be provided with an emergency stop but also have an automatic
slow-speed facility. In the Japanese and Russian standards
this is spelt out as 30 cm/s. The German and American drafts

prefer 25 cm/s but these speeds appear to be based on current
practice rather than any criterion which implies that speeds
below 25 are safe.

Safeguarding

All of the standards state the requirement of preventing access
to the robot when the automatic cycle is capable of being
initiated. In general this means a guard around the robot which
is interlocked to the robot cycle. In Germany, France and the
UK this usually means fencing 1.5 to 2 m high. In Japan a
lower standard seems to be accepted with single or double rail
barriers. The British approach points out that the type of
guarding depends on the application. They recommend that a risk
assessment is carried out and that as a result guarding may be
not required on, for example, small robots. The UK document also
emphasises the dangers of using software interlocking for guards
unless redundancy techniques are used or separate conventional
hardware interlocks.

User requirements

Both the American and British guidance outline the need for
safeworking procedures in addition to mechanical and electrical
safeguarding. These include the use of permits to work and
rules of access within the guarding for programming, teaching
and maintenance. The German draft also gives rules for robot
operation, both in automatic and setting-up modes.

Practical experience with industrial robots has shown that
programming and teaching is an important aspect of robot safety.
The need in many cases for the programmer to be close to the
robot arm, the possibilities of corruption of programs through
electrical interference or human carelessness can lead to
hazardous situations. Recommendations are given in the British,
American and German documents.

Little reference has been made so far to the Russian standard
which may have been based on the early Japanese and German
drafts. One interesting aspect here is a requirement for the
user to record and register any mishaps which occur. This, if

treated seriously, could be a useful inclusion in any international standard, as data on robot accidents and, in particular, near misses, is often difficult to obtain.

International standards

Work to develop an international standard for the safety of industrial robots has just started. At the first meeting in Frankfurt in May 1984, the International Standards Organisation (ISO) Working Group on robot safety, ISO/TC 184/SC2/WG3, considerably progress was made due largely to the similar approaches to the subject which have been taken by the individual countries. A framework for the standard has been agreed and comparisons made of the British, Japanese, German and American recommendations. These will be discussed at the next meeting of the Working Group in Paris in December and hopefully a first draft of the standard should be available in 1985. Initially the Working Group has limited its terms of reference to manipulatory industrial robots.

As stated earlier the safey aspects of programmable controllers when used with industrial robots has given some cause for concern. A Working Group of the International Electrotechnical Commission (IEC), IEC/TC44/WG1, has an interest in examining these problems within its wider brief to prepare standards for programmable controllers embracing machinery in general.

Conclusions

This paper has reviewed the current state-of-the-art in standards on the safety of robots. The high level of activity nationally and the progress at international level is a measure of the interest and concern being felt by robot manufacturers and users world-wide in the safety of their products. It is pleasing to note the unanimity in the approach of the individual countries and their willingness to cooperate at international level to ensure a safer working environment for all concerned.

References

1.   France   AFNOR standard - in preparation

2.   Germany (DDR)   TGL 30267/01, September 1982, Industrial
                       robots for machine tools; Terms;
                       Requirements of safeguarding measures

3.   Germany (FDR)   VDI Guideline 2853,   March 1984,
                       Safety requirements relating to the
                       construction, equipment and operation of
                       industrial robots and associated devices

4.   Japan   JIS B 8433-1983, General code for safety
                       of industrial robots

5.   Japan   Japan Industrial Safety and Health
                       Association, December 1983, Prevention of
                       industrial accidents due to industrial
                       robots

6.   United Kingdom   Machine Tool Trades Association, 1982,
                       Safeguarding industrial robots, Part 1:
                       Basic principles

7.   United Kingdom   Health and Safety Executive, Guidance
                       Note - in preparation

8.   USA   Robot Institute of America, draft, May
                       1984, Robot safety guidelines

9.   USSR   GOST-SSBT 12.2.072-82, Industrial robots,
                       robotised installations and robotised
                       shops.

Safety of Robot Installations in France

J. P. Vautrin
Laboratoire d'Electronique - Sécurité des Systemes
Institut National de Recherche et de Sécurité
F-54500 VANDOEUVRE lez NANCY

SUMMARY

The aim of this paper is to outline the current position on robot safety in France. It deals with robotics in relation to safety and working conditions, and considers enhanced safety resulting from robotics and new risks deriving from new methods.

Finally the paper summarises measures undertaken in France in the robot safety area.

1. GENERAL INFORMATION

There were more than 1300 industrial robots operating in France at the end of 1983. If up to now there has not been any serious accident due to the use of these industrial robots, manufacturers, users and authorities have been concerned with all aspects of safety in robotics, and various actions have been taken.

2. ROBOTICS, SAFETY AND WORKING CONDITIONS

In general terms robotics and automation of necessity have implications for health and safety. They have a major effect on production methods in manufacturing industry. These effects can be positive and negative. Although the main purpose of application of robotics is improved productivity there is no doubt that a transformation and therefore an improvement of working conditions is also possible. On the other hand robots by their very nature, or because of inadequate safeguards can become restrictive factors and even give rise to new hazards which have yet to be sufficiently taken into account.

2.1 Positive aspects

Promotion of safety

Robotics can allow of promotion of prevention of accidents and work-related illnesses.

. <u>Risks of mechanical origin</u>

Unlike semi-automatic machines such as presses, which require a human presence, industrial robots can function for a period without human intervention. This distancing of man from the machine and machining process is in itself a safety factor. For their safety, personnel must be kept out of the reach of mechanical danger either by obstacles at an appropriate distance from the machine or by deactivation of the machine. It is to be noted that these measures can only be easily applied while the machine is operating automatically. The processes of setting up and teaching the robot require a human presence in the immediate vicinity of the robot.

. <u>Other risks</u>

With regard to traditional machines, factors responsible for accidents or work-related illnesses can be classified into five groups :

- physical (other than mechanical risks)
- chemical
- biological
- physiological
- psychological

Looking at each group individually

a) the following are the physical factors :

- emission of particles or gases in the work area
- surface temperature of materials or equipment
- air temperature in the work area
- noise level at the work place
- level of vibration
- infrasonic vibration level
- ultrasonic vibration level
- air humidity
- speed of air replacement
- ionisation of the air
- level of ionising radiation in the work area
- voltage of electric circuits with direct or
      indirect risk of contact
- static electricity level
- level of electromagnetic radiation

- electrical field
- magnetic field
- lighting
- light intensity
- reduced contrast
- direct light reflection
- flickering of light
- level of infra-red radiation

b) chemical factors are linked to the presence of chemicals which can penetrate the human body through the respiratory system or eventually through the skin.

The following sub-groups can be listed with respect to their effects on the body :

- general poisons
- irritants
- sensitisors
- carcinogenics
- mutating factors influencing the reproductive function

c) Biological factors which can cause disease.

d) Physiological factors impinging largely on physical capacity :

- static overloading (posture)
- dynamic overleading (exertion, lifting or handling operations for example).

e) Psychological factors which under certain working conditions entail neuropsychological overloading.

Proper application of robots will prevent the factors listed above from having an undesirable effect on personnel in terms of accidents or work related illnesses. In France, as elsewhere, the tendency is to entrust to robots tasks which are disagreeable, routine, monotonous and dangerous such as spot welding, arc welding, flame cutting, spray painting and machine tool loading and unloading.

2.2 Risks inherant in a man - robot system

A robot is an automatic system which can undertake a given task without a human presence. Nevertheless, human intervention is not entirely dispensed with and is required during various phases of the robot life cycle. These are as follows :

- the first activation at the manufacturers
- programming directly by means of a syntaxer, or by manually leading the robot arm end
- adjusting
- maintenance
- repair

During all these phases man can be in intended or accidental contact with the main parts of the robot. The following is a non-exhaustive list of those likely to be in direct contact with robots - technicians of robot manufacturers, operators teaching robots; controllers and specialist technicians (e.g. in welding); supervisors; personnel working with other machines in the vicinity of the robot; inattentive employees. In the majority of cases a "man-robot" system of greater or lesser complexity is the order of the day with contact between man and the robot and its peripherals (arm, tool, workpiece). Thus as robots can operate with considerable force, and blindly, there is a permanent risk to man. In general, personnel can be protected during normal automatic functioning of the robot simply by forbidding access to the robotic operating area. However it is a different matter during other phases; particularly the teaching phase which often requires the presence of an operator in the vicinity of moving parts.

Risks specific to robotics

In comparison with traditional systems incorporating moving parts robots do not give rise to truly novel risks. Nevertheless the forceful, rapid (2m/s) and sometimes unforeseen movement of parts within a wider reach than is the case with more traditional automated equipment can create particular hazards. These can be classified as follows :

- collision between man and robot
- danger from projecting components
- wedging, jamming
- unforeseen commencement of motion
- other, such as electrocution, burns, types of radiation (*), electric arcs.

Risk sources (table A)

The risks can be categorised into two types :

those which involve the normal functioning of the robot and those which involve malfunctioning of the robot system. See Table A overleaf for a summary of these risks.

(*) e.g. - if the robot arm carries a laser source for a fettling or welding operation it will be obvious that an inopportune movement of the arm could cause grave risk to personnel even if at a considerable distance from the arm.

T A B L E   A

ROBOTS : RISK SOURCES

TYPE OF FUNCTIONS

RISK SOURCES	NORMAL	MALFUNCTIONS
More directly human	+ bad design of man/machine system	+ manual takeover gives rise to a potentially hazardous situation
	+ necessary entry into a non-authorised area	
	+ deliberate neutralising of safety mechanisms	
more directly technical	+ the guards fail to fulfil their safety function	+ failure of electronic control system (interface, software faults)
	+ the emergency stop is not sufficiently rapid for safety purposes	+ failure of electronic parts
		+ failure of hydraulic/ pneumatic components
	+ despite disconnection of energy sources the robot remains dangerous (e.g. potential energy, fall of arms)	+ failure of guards or emergency stop mechanisms
		+ various failures due to environmental factors (e.g. temperature, interference..)

3. ACTIONS IN HAND IN FRANCE IN THE AREA OF ROBOT SAFETY

Various actions have been undertaken in France, often at the initiative
of the INRS in order to ascertain the nature of the problems and the
required solutions.

3.1 Enquiry into the Problems of Safety of Robotic Installations

An enquiry is underway in this area involving co-operation between
engineers of the 'Services de Prévention de la Sécurité Sociale' and
industrialists. This incorporates a number of sections of which the most
important are the following :

- description of robotic applications
- improvement of working conditions and jobs
- potential risks resulting from the robot
- various specified malfunctions
    (types of failure -
       consequences of power cuts and mini-cuts)
- description of man-robot and machine-robot systems.

This enquiry is in progress and will be the subject of a publication.

Nevertheless some data can be provided at this stage. There has been one
accident in France involving a moulding machine-robot system (three weeks
out of action). The accident could have been avoided if access to the
system had been made impossible by a secure enclosure. It would anyway
appear to be obvious that a detailed study of methods of starting and
stopping automated systems would be of major importance.

3.2 Study of safety methods

Research is underway at INRS on the improvement of software and
hardware. Studies are underway on items of equipment such as presence
sensing mats, invisible barriers.

3.3 Standards

AFNOR is drafting a standard on robot safety. A draft will be
available at the end of 1984.

3.4 European Co-operation

France participates in meetings of the "Tripartite Group" on robot
safety with the FRG and UK and also in the work of "Robotics Europe".
France, through the media of INRS and SURF participates in the work co-
ordinated by the EEC entitled "Assessment of Programmable Electric Systems
with particular reference to Robotics".

CONCLUSION

From the above it will be seen that considerable progress is
being made in France in the matter of Robot Safety. Further,
France is interested in working with its European partners
through "Robotics Europe" to realise a common understanding.

Assessment of Programmable Electronic Systems with Particular Reference to Robotics

R BELL (Partner - CEC Collaborative Project)

Health & Safety Executive (FI5)
25 Chapel Street
London
NW1 5DT
United Kingdom

Synopsis

The Commission of the European Communities (CEC) is funding a joint project between seven organisations. These organisations already have programms of research in the field of Programmable Electronic Systems (PES's) and in particular robotic safety. The CEC funding provides the opportunity to collaborate in the exchange of information which will, hopefully, lead to harmonization of safety and reliability work in this field across Europe. Particular emphasis in the project is being placed on the provision of guidance on safety, reliability and availability assessments of PES's. This guidance will aim to ease the task of PES safety integrity assessment by considering the design, construction and manufacturing criteria, and the needs of the designer, user and regulatory authority.

INTRODUCTION

Computer based systems, generically referred to as Programmable Electronic Systems (PES's) are rapidly being deployed in a wide range of industries. Increasingly in the future there are likely to be pressures to use PES's for safety functions or for applications where there are safety implications in the event of PES failure. Unfortunately, the rapid evolution of PES's has meant that in the safety context the store of experience and advice, which in non-PES systems has taken many years to accumulate, is not available to the extent that has traditionally been the case. Yet if this new technology is to be effectively exploited it is essential that those responsible for making decisions in this area have sufficient guidance on the safety aspects on which to base those decisions. It is important that cognisance is taken of the work going on in other countries in order to minimise duplication of effort. It is also important to work towards agreement, between countries, of the broad safety principles involved. Agreement is more likely to be achieved at the present time because guidance on the safe use of PES's is still at a formulative stage.

CEC COLLABORATIVE PROJECT : BACKGROUND

On a European basis, the Commission of the European Communities (CEC) has agreed to fund a collaborative project between seven organisations. The participating organisations are, in the United Kingdom the Health and Safety Executive (HSE) who are the co-ordinating body, and the National Centre of Systems Reliability (NCSR); in Germany the Berufsgenossenschaftliches Institut fur Arbeitssicherheit (BIA) and the Fraunhofer Institut fur Produktionstechnik und Automatisierung (IPA); in Denmark the Elektronikcentralen (EC) and Arbejdstilsynet (At); and in France the Institut National de Recherche et de Securite (INRS). These organisations already have programmes of research in the field of PES's and in particular robotic safety. The funding from the CEC provides the opportunity to collaborate in the exchange of information which will, hopefully, lead to a harmonization of safety and reliability work in this field across Europe.

The collaborative project has identified a number of important objectives and milestones to be achieved during the two year project.

CEC COLLABORATIVE PROJECT : OBJECTIVES

There are seven Objectives which, apart from Objective 7 - the Seminar, are planned to be achieved within the two year period of the project. The Objectives are given below.

Objective 1 : Collection of information on PES's, specifically:-

   (a) Collection of safety, reliability and availability data.

(b)   Collection of accident and incident data.

(c)   Examination of classification schemes in existence for the collection of data on the field operation of PES's.

(d)   Analysis of the data obtained to provide a basis for the preparation of guidelines;

Objective 2 : Creation of a databank

The data obtained in Objective 1 will be classified and stored on NCSR's databank for analysis. This will provide a basis for setting reasonable and achievable standards for PES's and the determination of priorities within the project. The data will be made availabe in Europe at the end of the project.

Objective 3 : Collection and assesment of current guidelines, specifically:-

(a)   Collection of current guidelines for safe design, production, installation and operation of PES's. This would include standards and guidance documents produced by international and national standards making bodies, safety regulatory bodies, professional organisations, trade organisations etc.

(b)   Assessment of the guidelines, obtained in (a) above for their relevance in the context of robotic safety.

(c)   Collection and assessment of current and proposed work that is directly relevant to the formulation of guidelines.

Objective 4 : To identify, with appropriate justification, those areas where further work is required. This would identify the specific areas, in the context of robotic safety, where the PES plays a role in the overall safety, and would draw upon the results that have been obtained in the previous Objectives.

Objective 5 : This Objective will draw upon the work carried out in the previous Objectives and will seek to:-

(a) Formulate technical guidelines for the immediate future to provide guidance on safety, reliability and availability assessments of PES's including both hardware and software. Emphasis will be placed on quantitative techniques but it is recognized that in the immediate future qualitative techniques will have to be considered.

(b) Formulate future strategies for the use, development and review of the guidelines in (a) above.

Objective 6 : Promotion of the guidelines and results obtained in the collaborative project to achieve, specifically in the context of the guidelines, acceptance by regulatory bodies, international and national standards organisations, designers, manufacturers and user organisations.

Objective 7 : A seminar will be held at the termination of the collaborative project in order to publicise and aide the development of the results obtained. The overall management of the seminar will be the responsibility of the NCSR.

The aim of the project has been to structure the Objectives in a systematic manner so that the work involved with the latter Objectives are based upon the findings of the former. The main thrust of the project is concerned with PES's in general rather than concentrating solely on PES's associated with industrial robots although Objective 4 will, in fact, concentrate on the type of PES's used on industrial robots.

WORK PROGRAMME

The official start of the project was September 1983 and, as indicated above, the project is planned to last for two years. At the present time work has begun on all the Objectives except Objective 7. The amount of work done on some Objectives is more advanced than others due to the phasing of the work. The seminar is planned to be held six months after the official termination of the project.

CEC COLLABORATIVE PROJECT : RELEVANCE TO INDUSTRIAL ROBOTS

When assessing the adequacy of the safeguards provided on an industrial robot application, it may be necessary to consider, the safety integrity of the industrial robot control system. In many cases it may be possible to adopt a safeguarding strategy which is independent of the control system eg by hardwired safety interlocks, physical barriers, mechanical stops to limit the arc of movement, by speed and torque limitation arranged by non-PES devices and by well defined and regulated systems of work. From a safety assessment view point there is a lot to be gained by the adoption of safeguards that can be readily assessed and on which there is established guidance. Nevertheless, there are situations where personal safety, and minimisation of robot damage from adjacent structures and plant, is dependent upon the continuing correct functioning of the robot control system. In some cases there may be economic penalties by adopting a safeguarding strategy independent of the control system. Whilst, as indicated above, there are many advantages in using conventional safety devices, and this is to be strongly recommended in many cases, it would be unreasonable to prevent in principle the use of PES technology for safety functions. An important aim of the project is to develop guidance that can be used for the assessment of the safety integrity of robot control systems incorporating PES's.

The project is complimentary to the work being done in IRSE and the Tripartite Working Group on Robotic Safety in which France, West Germany and the United Kingdom are involved. It is important that close liaison between these groups is undertaken in order to obtain the maximum benefit of international co-operation and in order to avoid any duplication of effort.

WAY-AHEAD

For the work in the collaborative project to be fully utilised in practice, it is essential that the various groups involved in robotic safety have a clear understanding of each others aims and areas of work. In this context, there is a need for effective liaison between the Collaborative Project, dealing specifically with the assessment of PES's, and on the other hand, the two groups dealing with the overall safety issues raised by the use of industrial robots - namely, IRSE and the Tripartite Working Group on Robotic Safety. Cognisance must also be taken of the work being done in the international standards organisations - particularly ISO TC184, IEC 44 and IEC 65.

If the work being done under the auspices of the CEC, both in the context of the assessment of PES's and overall safety of robotic installations, is to be effective in both the short and long term it is necessary to ensure that the ideas and views being developed, which are still at the formulative stage, are fed into the various national safety regulatory authorities and national and international standards making bodies. Only if this is achieved will the funding by the CEC have a lasting benefit to Europe.

Occupational Safety and Industrial Robots – Present Stage of Discussions within the Tripartite Group on Robotic Safety

PETER NICOLAISEN

Fraunhofer-Institut für Produktionstechnik
und Automatisierung (IPA), Stuttgart
German Federal Republic

Summary

Often, when industrial robots have been in use, not enough
attention has been paid to occupational safety. On the other
hand, it is very difficult for many users to find a suitable
solution for their problem. Starting with a representation of
the problem spectrum as a whole, starting points are indicated and discussed for the improvement of safety at workplaces
where industrial robots are used.

1.   Introduction

The subject of 'industrial robots' occupies a position in pub-
lic discussion which lies far above its real position in the
economy. The main discussion points are, however, technical
problems and aspects relating to development and use, supported or thrown into doubt by socio-economic arguments such as
'Industrial robots free men from hard, energetic dangerous
work' (Keyword:humanisation of working life) or 'Industrial
robots eliminate employment and lead to other activities de-
manding a low level of qualification' (Keyword: job killer).

What gradually emerges is an awareness of the problem with regard to the accident   associated with the use of industrial robots. Here too, as the newspaper report in Fig. 1 demonstrates, there is a disproportionate over-reaction, probably
because of disappointment that the apparent 'jack of all trades'
industrial robot often highly stylised to appear like a mechanical man, is not perfect and can even turn against 'its master'.
Yet to make this subject more accessible to analysis and solution, it would require to be examined in a more sober fashion.

"STRUCK DOWN BY A ROBOT"

Tokio (AP). In Japan, for the first time a man was murdered by a robot. According to the report of an enquiry the incident occurred way back in June, in the Tokyo firm of Kawasaki, but was published only after enquiries were complete. The robot's victim was a 37 year-old maintenance worker, whom the report accused of carelessness. It was also stated that the safety arrangements in the factory were inadequate.

According to the enquiry report, the maintenance worker had crossed a safety barrier in order to work on a defective machine with which the automobile gears were made. In doing so, he accidentally set a robot in motion; its arm moved forward, pushed the worker in the back and pressed him against the machine. The worker died of internal injuries in hospital.

In Japan, about 70,000 robots are used in industrial production processes.

Fig. 1: Newspaper cutting on an occupational accident in which an industrial robot was concerned.

2.   Problem

2.1   Accident risks   specific to industrial robots

Accident hazards due to industrial robots arise mainly because of the h i g h - p o w e r   m o v e m e n t s   wide area covered by movement, high speed of displacement, large masses), which too, in contrast to conventional machines are f r e e l y p r o g r a m m a b l e as regards r o u t e   a n d   s p e e d o f   m o v e m e n t (Fig. 2).

INDUSTRIAL ROBOTS	CONVENTIONAL MACHINES
Simultaneous movement in several (up to 8) axes	Usually only simultaneous movement in few (1-2) axes
Free programmability of the speeds of every seperate axes	Pre-set fixed speed
Free programmability of direction of movement of every separate axes (free spacial movement)	Fixed movement pattern (pre-set routes)
Very large range of movement compared with the volume of the appliances	Range of movement usually smaller than volume of machine
Range of movement overlaps the position of other machines, parts of buildings etc.	Scarcely any overlapping

<u>Fig. 2:</u> Comparison of industrial robots and conventional machines

Even in the normal case it is hardly possible for an outsider to think ahead to the next movement, so it becomes still more problematic, when, because of a disturbance (e.g. in the positioning control or in the speed monitoring) completely unpredictable movements, with undefined speed, occur within the (kinematically possible) range of movement.

In contrast to conventional machines, in which a movement takes place usually within the machine, in the case of industrial robots, the size and shape of the danger area i.e. of the relevant working space and maximum range of movement, respectively, is not immediately recognisable. For this reason, precautions must be taken to safeguard this zone.

2.2   The accident event

Everyone who has dealt with industrial robots for a fairly long
time will be aware of the fact that the topic 'Occupational
safety and industrial robots' is not of purely academic interest,
but has a real practical basis; everyone can remember critical
situations in which only a lucky chance has prevented a 'near'
accident becoming an 'actual' accident.

It is, however, much more difficult to get statistics on accidents. From most of the countries using a large number of industrial robots, either no data or only vague data are obtain-
able and the percentage of unrecorded accidents might presumab-
ly be high)Fig.3). Only Sweden forms an exception in this respect.

ACCIDENTS WITH INDUSTRIAL ROBOTS	
COUNTRY	REPORTED ACCIDENTS
GERMANY	According to the Trade Assurance Assoc. no accidents with IRs have been reported. (but there have been several accidents)
FRANCE	Several accidents
JAPAN	4 dead Several injured
SWEDEN	Accident figures from an IR user reported (for the period 1977-1981, the number of IRs in use being 44). 17 accidents involving injuries to persons 3 of these accidents caused serious injuries. About 5o more accidents reported from other companies. 1 dead
BELGIUM	Accident with serious injury
FINLAND	2 accidents reported

Fig. 3: Accidents at workplaces where robots are
         in use (as far as has been reported).

From Sweden also came the results (quoted in Fig.4) of a 14-
day survey of eight IR workplaces. In the course of this inve-
stigation dating from 1981, a total of 24 critical situations
were recorded, the cases indicated by a black point o leading
to injuries or material damage.

Dangerous situation observed	Cause	Pointers for solving the problem
1 IR brushes against operator	Control panel is within the enclosure, operator passes through the zone where the IR is moving	LAYOUT
2 IR brushes against person	No all-round enclosure; there is a gap between two machines	LAYOUT
3 IR brushes against person/ operator/machine-setter du- ring his work	Parameters of machine are adjusted and material fed in while the installation is running	LAYOUT ORGANISATION OF WORK
4 Operator is struck by the IR	Sometimes the workpiece re- mains hanging on the conve- yor belt and the operator intervenes while the instal- lation is running	WORK ORGANISAT. LAYOUT WORKING PROCESS
5 Operator is struck by IR while he is working	Range of activity (workpiece being put from bunker into store) is partly outside the working zone of the IR	LAYOUT WORK ORGANISAT.
6 IR projects the workpiece at high speed	No pressure or too low pres- sure for pneumatic gripper	DESIGN LAYOUT
7 IR brushes against person (toolsetter)	Pneumatic cylinders have to be adjusted with the instal- lation in operation	LAYOUT WORK ORGANISAT.
8 Danger of being burnt by HF installation	There are various emergency cut-out circuits. If the emergency cut-out circuit is released the HF preheating system starts automatically	LAYOUT (INTERLINKING)
9 Person is struck by arm of IR or by workpiece	Safety grid fitted wrongly and is too low. IR reaches out over the fence	LAYOUT
10 Person is struck by IR while he is working and can be burnt by fluid	Range of activity (feeding in liquid metal) is partly within the programmed move- ment of the IR	LAYOUT WORK ORGANISAT.
11 Person is struck by IR, dan- ger of burns from liquid spray	Remains of cast metal are usually removed from the workpiece while the instal- lation is running	LAYOUT WORK ORGANISAT. WORKING PROCESS
12 An inquisitive person is struck by IR, danger of bur- ning from hot workpiece	Gaps in fencing	LAYOUT
13 Hand of operator gets jammed between IR and barrier o Injury (7 days absence from work)	Too short a distance between IR working zone and barrier. Barriers mesh is too wide and /or barrier is too low	LAYOUT

Fig. 4: Accident hazards with the use of industrial robots -
Results of an observation of 8 workplaces over a

Dangerous situation observed	Cause	Pointers for solving the problem
14 Programmer is struck by IR although he is acting correctly	When, during a MANUAL type operation (programming) the safety barrier is opened, the IR does not stop immediately but completes its cycle	LAYOUT INTERLINKING
15 Person is struck by IR	Safety barrier has gaps because after a machine has been replaced the old (now too small) barrier is still in use	LAYOUT WORK ORGANISAT.
16 Person is struck by IR	Paletting system for workpieces is part of the safety enclosure; when palettes are removed there is free access	LAYOUT
17 Operator is struck by IR	Operator, in the course of his work, (lifting the workpiece and inserting it) bend into the operation zone of the IR	LAYOUT WORK ORGANISAT.
18 Slipping on oily floor	Because the spray nozzle has been wrongly positioned oil runs over the IR axis and onto the floor behind	WORKING PROCESS
19 Operator's hand is jammed between the IR and hot workpiece ● Burn injury	The amount of oil sprayed by the IR is sometimes insufficient (because of faulty nozzle setting) so that when the plant is running more oil has to be added, by hand	WORKING PROCESS
20 Operator is struck by blown-out metal pieces (trimming press) (●) Cuts due to metal splinters but no actual accident involving IR	Different reasons: Position and pressure of blow-out nozzles, position of operator	LAYOUT WORK ORGANISAT. WORKING PROCESS
21 Stumbling over cable between IR and control panel ● Injury (bruising)	Cable not covered and badly laid	LAYOUT WORK ORGANISAT.
22 Maintenance worker is pressed by the IR against a running grinding wheel ● Cuts and burns	Maintenance worker had stopped the plant via the emergency cut-out but did not know that the emergency cut-out only stops the IR (hydraulics out). When, in cleaning, he inadvertently touches a limit switch, the grinding machine starts up and the IR, without pressure, caves in and presses the operators arm against the moving grinding machines	DESIGN LAYOUT INTERLINKING
23 Risk of injury by projected workpiece	The feed line of a compressed-air operated gripping tool is broken off so that the workpiece can no longer be held	DESIGN LAYOUT
24 IR makes uncontrolled movements and collides with machine, whereby the gripping tool is bent ●	A lead was broken by dangling workpieces and passed across the various control signals (including the emergency cut-out). This caused the entire plant, including the IR to get out of control	DESIGN LAYOUT INTERLINKING

Fig.4: continued

2.3   Future trends

The problem of accident hazards as a whole is not visible at a
glance if only for the fact that, so far, industrial robots are
used in relatively few fields and this, as a rule, in large production
batch(flexibility hardly used), so that the technical advantages, which, at the same time, form their special danger, are
often not exploited.

In the future, however, this situation will alter more and more
in certain circumstances, this will cause an increased accident
hazard, when:

- owing to the wider use of robots in small and medium size
  batch production the amount   of programming work to
  be carried out will increase (more frequent contact between
  the person and the industrial robot).

- the range of functions will be extended so that the manufacturing systems as a whole become more complex and thus, in
  certain circumstances become more prone to breakdowns, i.e.,

  o more complex programmes
  o collaboration of several industrial robots
  o use of sensors (external data)
  o use of gripper-/tool-changing systems
  o mobile robots (e.g. mounted on vehicles)

- the energy potential available within the system increases

  o industrial robots with higher performance
  o use of high-speed tools
  o laser and water-jet cutting with industrial robots

2.4   Present state of safety technology

In spite of some differently orientated problems, it has, so
far, been a question of conventional safety technology. The
safety fence frequently designed, so that, when it is opened,

the installation automatically stops, is practically the only
appropriate safety device.
The reasons for this are obvious:
- usually simple to produce
- inexpensive protection outside the fencing
- protection against 'thrown-away' parts

So, in the near future, these safety devices will be resorted
to again. It is true that the safety fence is no universal pana-
cea: on the one hand, there are plants where it cannot be installed for reasons of the manufacturing process being carried
out and, on the other hand it affords no protection when work
has to be carried out inside the enclosure (fitting, programming, maintenance, inspection, repair).

Fig. 5 shows typical safety equipment for an IR (industrial
robot) workplace. In practice, however, there is often a short-
age of safety devices, as the examples shown in Fig. 6 (typical of many others) demonstrate.

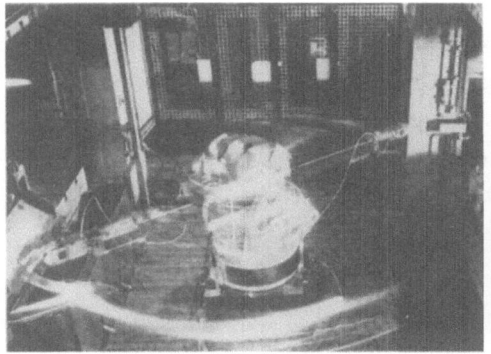

Fig. 5: Typical safety arrangements for an IR workplace
(from VW)

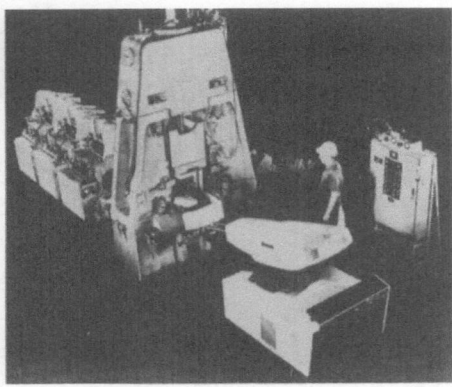

Fig. 6: Examples of IR workplaces with inadequate safety
        devices

Although it might be simple, from the legal point of view, to
indicate the 'responsible'and the 'guilty' parties, it is very
difficult and timeconsuming to improve matters. For it is not
(always) the 'blinkered vision and egoism of rabid technocrats'
or a 'misanthropic capitalistic way of thinking' which causes
this deficiency but, more often it is an initial unawareness
of the problem and a certain perplexity in seeking a solution,
which can be explained by several factors:

- in the development and introduction of new technologies it is
  usually the question of technology and sales policy that are
  in the forefront and safety problems have a lower priority.

- awareness of a problem often develops after several negative
  experiences i.e. on the basis of a large number of applications and after years of use.

- assessment of the problem (as to what level of safety is
  necessary) varies according to the interests of the manufacturer, user and safety authorities.

- solution of the problem raises difficulties because specially prepared information material, relevant aids such as check-lists, catalogues, collections of examples and suitable safety devices are lacking.

3. Pointers for the improvement of occupational safety in the use of industrial robots

From the foregoing remarks it will be seen that there are two important sets of factors involved, when it is a question of increasing safety at workplaces where industrial robots are in use.

In order to achieve success, it is essential to make all concerned aware of the problem. For it is only when the problem is detected and recognised as such, will safety problems be given the same priority as questions of technology or economy, so that it is self-evident that they should be included in the considerations at a very early stage (a great step forward has recently been made with regard to the latter point).

Yet all the insight and motivation is of no help if they are not followed up by parallel activities, dealing with the solution of the problem, i.e.

- proposing methods of finding a solution

- putting forward and evaluating principles and examples

- proceeding with further and new developments, in order to close gaps or eliminate faults

Given a working party which includes a wide range of specialisms- such as is found within the British-German-French Tripartite Group - we consider it sensible to bring together and to discuss the experience gained so far so as to arrive at solutions in a constructive manner; solutions which, on the one hand, meet the requirements of various groups (such as

constructors, industrial planners, plant operators, safety
specialtists, legislators) whilst taking account of various
national idiosyncracies (in the area, broadly speaking, of
accident prevention legislation), and on the other hand taking care not to have everyone involved 'doing his own thing'
and creating what in some situations could be insuperable
difficulties for his neighbours.

The discussion on the question of the approach and method of
presentation which would be most appropriate as an aid to decision making in the matter of how 'industrial robot workplaces' can be designed and operated for safety gave the following results (Fig. 7).

The flow chart shown is the basis for this process, and has
been presented in the general form which it has in order that
it will be useful in the most diverse types of problem situations, areas of application and phases of planning.

For the various stages in the 'workplace-with-individual-robot'
system, such as transport, installation/linkage, programming/
arranging etc., hazard analyses are carried out (corresponding
with varying degrees of precision, to the planning phase). The
aim of these is to determine the existing dangers so that coun-
termeasures can be taken.

The optimum moment for a first run-through of the flow chart is
generally at a very early stage for, often by fixing of the
plan or design we (unknowingly) take a decision as to what kind
of safety can be achieved for the system as a whole (e.g. what
type of control; drives with/without additional brake).

As, in practice, however, it can not be assumed that there have
already been detailed discussions with experts from the occupational safety department (however desirable that would be), it
is all the more important to acquaint the constructor and the
planner of the plant with the fundamental ideas of occupational
safety.

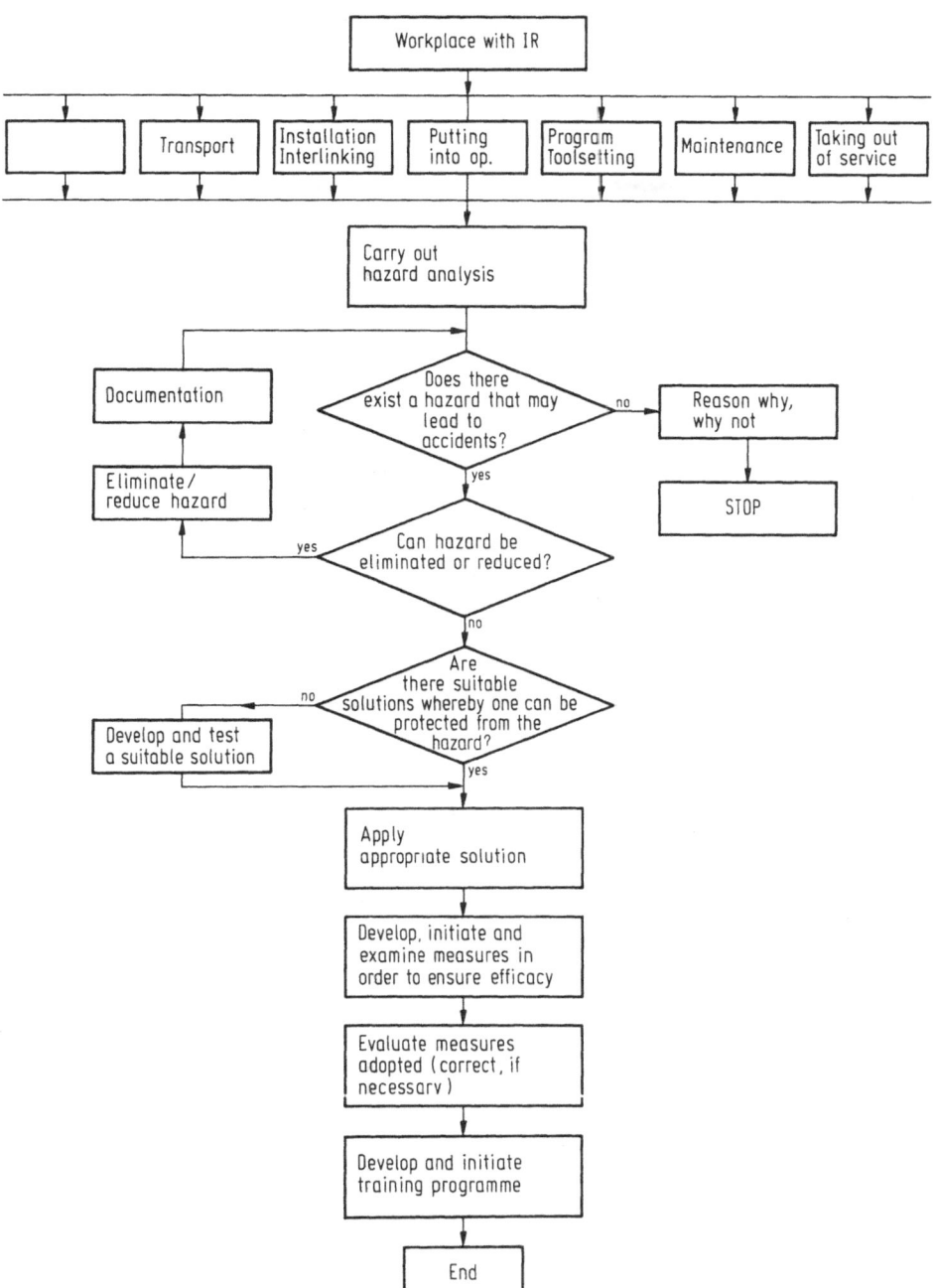

Fig. 7: Diagram showing different stages in the accident prevention procedure

The immediate task for our Tripartite Group, however, consists on providing a collection of material and data (Fig. 8) which will enable

- the different groups of users to be supplied with a range of relevant basic information

- extensions and modifications to be carried out (say, when new problems arise or when new safety regulations are brought out)

- specific operational data or special national rules and regulations to be incorporated/integrated without wrecking the overall concept

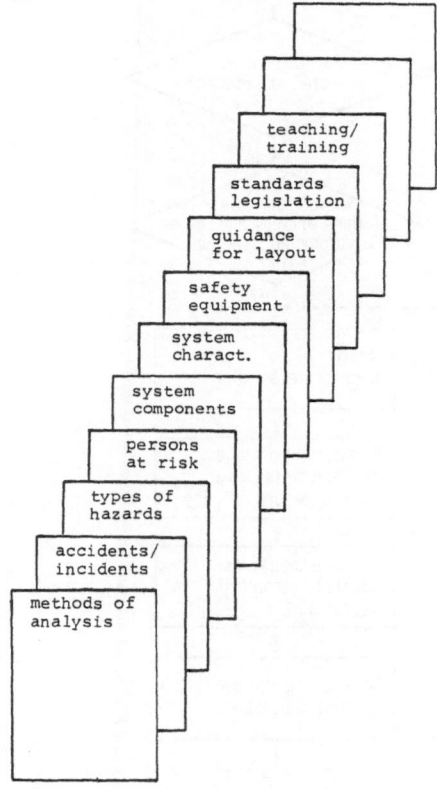

Fig. 8: Breakdown of material collected on the subject of 'occupational safety when using industrial robots'

Only as a long-term target we see the development of new standards and guidelines and/or modification and extension of existing ones.

This task must at the present time at least be considered to be beyond the scope of this group.

Even at a future stage it will not be possible to run through the flow chart quasi-automatically. Now as ever it will be necessary to devote a lot of ones own thought to defining, delimiting and solving the problem. What can be done, however, to make the work easier and more efficient by drawing on the experiences of others by way of checklists, forms, catalogues and sample solutions. Meantime, there will be a lot of activity in this direction i.e. collecting information, exchanging experiences (project group on material handling systems (ARGE HHS), VDI, DIN, Automotive Association, ISO, IEC, EEC (WGS), Tripartite Group), and although, nearly everywhere, one is still at the beginning and only isolated, preliminary results are available it is urged that the work should be coordinated, both in order to avoid duplicating the work and in order to prevent each group working for itself and trying to impose its own method and thus building up insurmountable obstacles for its neighbours. The Tripartite Group has made this coordination its special concern.

Although the collection of data and material on robotic safety has made progress, it is still a long way to go for its completion and to list it up in a way to be a real support for the design engineer, the layout planner or the safety engineer.

At the same time have to be treated further main topics, which have hardly been studied until now. On the one hand, the task is

- to develop rules and guidelines for the design. This is a wide field of activity because of the great number of possible factors of influence. The necessity to work out proposals for a safety-oriented layout-design results on the one hand on the analysis of accidents that have already occurred (more than 50% of all accidents are due to a large extent to an inadequate layout-design), in the other hand on the knowledge that this problem will still exist even if tremendous progress is achieved on other fields e.g. in the field of safety-oriented design of industrial robots. (A safe robot does not ensure a safe plant even if the robot represents an important factor of influence).

- In the field of safety devices for IR-workplaces there is still a need of support for the inexperienced user to deal with his safety problem. A catalogue containing all available safety devices has to be compiled, completed by a collection of examples of all realized solutions. On the other hand it is necessary to improve existing or to develop new safety devices to achieve better solutions of the problem. A first step in this direction represents the above mentioned large-area switch developed at IPA for use at IR-workplaces, but besides this, there is still a wide range of possible applications in industry (e.g. guided vehicles, moving parts of machines). This device offers effective protection especially for operations in the immediate vicinity of the industrial robot as it is fitted directly on the apparatus and travels with it (Fig. 9).

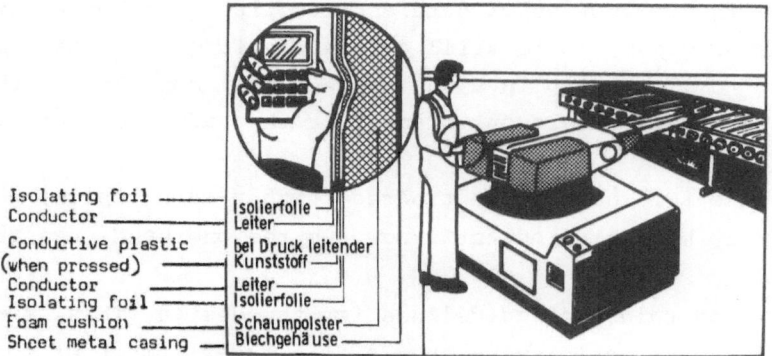

Isolating foil — Isolierfolie
Conductor — Leiter
Conductive plastic — bei Druck leitender
(when pressed) — Kunststoff
Conductor — Leiter
Isolating foil — Isolierfolie
Foam cushion — Schaumpolster
Sheet metal casing — Blechgehäuse

Fig. 9: Principle of the safety switch

It consists, essentially, of two components:

- a large-area switch, which when an obstacle (person, machine, support) is encountered, gives a signal which stops
  the movement

- a suitable structure which, like the dashboard in a car can
  change on impact and thus reaches the motional energy still
  remaining after the disconnection.

The signals given by the large-area switch are evaluated by
means of an electronic circuit, which according to the requirements for safety devices of the German Safety Authorities, controls faults as breakage of cables or short-circuit and helps
to maintain the safe function of the robot in case of a breakdown of components.

Conclusion

Everyone who has dealt with industrial robots will know that
a solution to the problem 'Occupational safety and industrial
robots' cannot be achieved by the realization of one single
'great idea', but only by the more or less successful combination of different partial solutions worked out by the project teams.
To ameliorate and facilitate this cooperation will be the most
important task for the future.

Towards Developing Reliability and Safety Related Standards Using Systematic Methodologies

F. DUGGAN, R. H. JONES and K. KHODABANDEHLOO

Centre for Robotics and Automated Systems
Department of Mechanical Engineering
Imperial College
London   SW7 2BX.
United Kingdom

Abstract
Aspects of Robot safety have been considered using logical reliability assessment techniques. Fault Tree Analysis has been used to assess the reliability of the robot itself, the interaction of the robot and other machinery excluding time dependences and on a descriptive basis the interaction of the robot with humans. Event Tree Analysis has been used to examine the hazards that may be encountered in a cell and to assess the adequacy of the compensating safety features. All these aspects have been illustrated with simple examples indicating in brief the scope and possibilities of developing systematic methodies for safety related requirements and standards. This paper is addressed to methodologies and no attempt has been made to define any standards.

1. INTRODUCTION

The need for safety related standards is obvious, but there is no simple way of deriving such standards. This is hardly suprising as safety is a product of many separate elements. For instance reliability of the robot and allied equipment, maintenance of all equipment, the integrity of the software, human errors during operation are a few contributing effects that need to be considered. Such factors require a detailed study both on their own and as an integral part of the robotic system. Here we take a much less ambitious viewpoint. We consider a reliability and safety analysis of the system as a pragmatic approach for developing reliable and safe robotic systems. In this way we can draw upon logical reliability and safety methodologies [1, 2] that have previously proved successful. By considering such methodologies in relation to robotic systems it is possible to identify potentially dangerous situations for which safety measures need to be adopted. Traditionally a mixture of good engineering design, judgement and experience is used along with good maintenance to achieve a safe, reliable system. Furthermore, any particularly dangerous situations which endanger personnel are safe-guarded by fencing. This method of guarding does place a barrier between the robot and the human while the robot is in playback mode. However, there are other human-machine interactions, which are not guarded by fencing notably maintenance and robot teaching. It is precisely these sort of events that call for a systematic approach to indicate potentially dangerous situations. Once hazards have been identified it is possible to consider safety related requirements and standards. In this paper we only consider methodologies and do not define standards. In sections 2 and 3 we show how Fault Tree Analysis (FTA) may be used to asses the reliability of the robot itself and how robot-machine and robot-human interactions may be examined. In section 4 Event Tree Analysis (ETA) is used to assess the hazards that may be encountered in a robotic cell. In section 5 some concluding remarks are made.

2.   ASSESSMENT OF THE ROBOT ITSELF

The reliability of hardware can be assessed using the logical reliability methodology of Fault Tree analysis [1, 2] and this method has been applied to an industrial robot [3]. Before FTA could be applied the design details of the robot or subsytems of the robot are needed. A Failure Mode and Effect Analysis (FMEA) is carried out as indicated elsewhere [3]. In FMEA the failure modes of each component are assessed, from which potentially major failure consequences are identified. In FTA terminology these major consequences are called Top Events. Here we study the top event "unwanted arm movement in teach mode" for a hydraulic robot shown in Fig. 1, which was also considered in Ref. 3. The fault tree for this top event is presented in Fig. 2 from which the following points emerge:

(a) Any one of three categories of events can individually cause the top event: these being servo, drive or teach pendant failures. All these failures can be further traced through the system to failures called basic events to which a failure probability can be assigned independently.

(b) Drive failures are dependent both on transmission failure and on hydraulic drive failure. A typical example of drive failure has been adequately presented in ref. 3 and so will not be developed further in this paper. However, other failure branches of fig. 2 will be followed through.

(c) Events like the servo power amplifier failure can be further developed but may be treated as a primary event to which a failure probability may be assigned.

(d) The hydraulic pressure supply failure leads to a sub-tree given in fig. 3. The analysis of such trees could contribute to the assessments required for the preparation of check lists. On a more quantitative basis the maximum permissible failure rate for seals, valves and other equipment can be arrived at after evaluating [1, 2] the tree failure probabilities.

(e) The sub-tree for DC power supply failure causing unwanted movement is given in fig. 4 and may be interpreted in the same way as fig. 3. Further use will be made of this tree in section 3.

(f) In fig. 5, the teach pendant failure sub-tree is presented. An unwanted movement can be caused when the trigger button fails in closed mode and any one of the joint buttons also fails in a closed mode.

From the above qualitative analysis using fault trees given in figs. 2-5, it is clear that failures causing unwanted movement during robot teaching can be identified. However, the relative importance of the contributing failures can be obtained after quantitatively evaluating [1, 2] such fault trees, which are presently under way. From such quantitative information the critical component failures can be guarded against by prescribing standards. Such standards could be in the form of any one or combination of measures as appropriate: the maximum allowable probability of occurence of a given dangerous top event; the specification of the minimum quality of certain critical components; and additional levels of compensating safety features to safeguard against crucial component failures can be required of the robot designs. As a by product on a more qualitative basis FTA can help in drawing up checklists as a possible means of communicating design and system related information. Finally it should be noted that in the above example we have only considered random component failures but other sources of failures can be included to refine the analysis, some of which have been considered in the next section.

3. ANALYSIS OF ROBOT-MACHINE AND ROBOT-HUMAN INTERACTIONS

Fault tree analysis can also be extended to consider failures arising from interactions with the robot like that of other machinery and people. In such a case the fault tree of fig. 2 is modified to that of fig. 6, which is only partially developed, but serves the purpose of illustrating such an analysis.

To consider robot-machine and robot human interactions, the entire work place design would have to be considered in detail including safety interlocks. Rather than develop further these two new branches of fig. 6, we illustrate such interactions by referring to the earlier fault trees.

An example of external sources causing robot movement is implicit in fig. 4, where mains noise of sufficient magnitude can induce a spurious movement of the robot. Such a movement would be unsafe during teach mode both to the operator and to the hardware, but in normal operational mode it could only cause damage to the robot and/or other machinery. Fault trees can be extended to consider interactions of the robot with other hardware for which failure rates are readily available. However, time dependencies would not be taken into acount. Such a large system would produce vast trees, which can be mitigated by subdividing the trees into modules.

In contrast human errors are not easily quantified or well understood and so present difficulties. However the inclusion of human error branches in fault trees is possible and in some cases illuminating as it can help in increasing awareness of the system dependence on human interaction with the system. Such fault trees can aid in the development of training programmes, maintenance schedules and practices. Although quantification of human errors on a probablistic basis is difficult if at all possible, a realistic band of probabilities can be used so as to identify the sensitivity of the top event probability to human errors. In some cases where human error combines with other system failures through a logical OR gate, the top event becomes dominated by human error probability. This implies that additional design features are essential to eliminate, at least in part, such dependancies.

The teach pendant failure of fig. 5 is modified to include human errors and is presented in fig. 7. The branches where both the trigger and any one of the joint buttons fail correspond to the case considered in fig. 5. The other three branches involve human action: if the trigger button fails closed and the operator presses any of the joint buttons; similarly if one of the joint buttons fails closed and the operator presses the trigger; or finally if the operator inadvertently presses both the trigger button and any one of the joint buttons. Even from the qualitative tree of fig. 7 it is clear that standards both for the failure rate of equipment and ease of correct use of the teach pendant could be considered. If it is found that pendant failure contributes significantly to the top event, or if the design layout is found to increase the risk of inadvertent button pressing, redesign could also be considered as the means of hazard reduction.

4. ANALYSIS OF A ROBOTIC CELL

An alternative reliability analysis to FTA for hazard assessment is Event Tree Analysis [1, 2]. This has been applied to a robot welding cell [4]. In order to apply ETA the design details of the robotic cell are needed. Event trees yield very large trees and so the ETA is considered at subsystem level. However, effects at a more detailed level like component failure are implicit in the ETA.

In order to illustrate the technique of ETA we consider a simple robot welding system [4] shown in fig. 8. An event tree is presented in fig. 9 for the frequently occuring event of the operator entering the working area for adjustment or programming. In ETA terminology such an event is called an initiating event.

As shown in fig. 8 the robot cell is guarded by fencing. Inside the fencing is a microprocessor controlled multijointed robot, its teach pendant incorporating an emergency stop, a welding unit and a two position turntable. Access to the cell is allowed only through a gate. In normal production the access gate is shut and the two position turntable is loaded by the operator from the working area. Outside the fencing there is a start button and a system emergency stop, which stops all equipment. The turntable and robot consoles are also located outside the fencing. Such a cell is designed with care using engineering experience. This can be augmented by ETA to check that all hazards have been adequately considered and that the cell is safe. These activities should be considered at the stage of cell design.

In fig. 8 the work spaces of both the robot and the turntable are marked. For the event tree of fig. 9, it is assumed that the operator has entered the cell and is within either or both of these workspaces. For such an initiating event there are several hazards from failures and wrong action, which the operator might encounter. These events are listed across the top of the tree. Each of these events may or may not occur which is shown in fig. 9 by Y or N respectively. This gives rise to the branches of the event tree. The order in which these events are arranged usually has to be in the time sequence of occurring events. When the events are time independent the order can be used to minimise the number of branches. For instance in this example the natural position for the system emergency stop is next to the initiating event because once this emergency stop is activated all equipment is inoperative. However, this is not the case for the teach pendant emergency stop, which is capable of stopping all robot movements and so it is convenient to place this before any other robot related event. Finally it should be noted that the crucially important feature of ETA is in the choice of these events as their proper choice can highlight problem areas, but if badly chosen it can obscure them.

The event tree of fig. 9 has 39 separate final branches that may occur due to the eight events considered. These 39 sequences of events that may occur are conveniently classified into six outcomes of varying degrees of hazards (state numbers 2, 3, 5, 6, 7, 8) and two safe outcomes (state numbers 1 and 4). Clearly such a logical analysis of the safety of any system is illuminating and useful for setting standards of compensating safety systems. It can also be used to make the operators more aware of the potential hazards they face.

From considering fig. 9 the following points emerge.
(a) Considering the safe system outcomes i.e. state numbers 1 and 4: We find that the system is inactive in state 1, if either the system emergency stop is activated or if the turntable and weld gun are switched off and the teach pendant emergency stop is activated. The system is still safe in state 4 and the operator can perform the task satisfactorily, if the turntable and weld gun are switched off, and no unexpected robot movement occurs. Thus an ETA identifies the circumstances in which unexpected robot movement could be dangerous. Such events ascertained from ETA can be incorporated into a robotic design standard stating the requirement that the probability of the occurrence of unwanted movement never exceeds a given value. For any design of robot this can be specified.
(b) Furthermore an examination of states 2 and 3 again highlights the need for avoiding unexpected robot movement. However, these states do not necessarily point to the unreliability of the robot, but rather to the need for a d d i t i o n a l compensating safety systems to obviate such an outcome in teach mode. For instance the operational speed of the robot in this mode could be reduced by additional controlling hardware. Such restricters could be automatically removed in ordinary operational circumstances. It should be noted that if any limiters were software based it cannot be relied upon as a safety feature, but only considered as an a d d i t i o n a l compensating safety feature

after ensuring that the design of the robot is as reliable as possible.

(c) The states 6 and 8 in effect amount to not having the turntable and weld gun switched off. The situation could be remedied by good training and/or proper interlocking.

(d) States 5 and 7 caused by unexpected behaviour of weld gun and turntable can be treated in the same way as that of the robot.

(e) The multiple failure states can be reduced in intensity by ensuring good training and proper practices as long as the basic equipment is reliable and the safety systems design is adequate.

In the above example we have considered a cell design and used ETA to assess the hazards and prescribe compensating safety features. From a consideration of the event tree of fig. 9 it does appear that the following safety features ought to be incorporated.

(i)   Assuming that the robot is reliable it would be useful to add hardware limiters in teach mode to avoid fast robot movements.

(ii)   In order to ensure that the turntable and weld gun are switched off, an automatic cut off interlocked with the access door would be needed. However, an override box would be needed so that the operator may activate these pieces of equipment as they are often required while teaching. The robot would also need to be interlocked with the access gate in the same way and an override switch for it would also need to be provided to allow the robot to be taught its programmed path. However, the gate interlock is not the same as a system emergency stop since it would operate though the I/O ports of the machinery in order to avoid loss of running programmes.

(iii) When the access gate is open the weld gun and turntable are switched off as above. However, if the override switches are used to activate them during teaching, then the robot controllers must act as the system supervisor, which was not immediately obvious prior to the ETA. Using the robot controller as a system supervisor has two consequences. If an emergency does arise the operator can control all the equipment from the teach pendant's emergency stop. Nevertheless spurious signals from the robot could cause unexpected behaviour of the turntable and/or the weld gun. This must be guarded against at least by proper operator awareness and training.

All teaching activities could be carried out at low speed, but for arc welding it is sometimes necessary to test the program at working speed. For such a test an operator would be required to remain inside the cell within easy access of the teach pendant's emergency stop. Another operator would be outside the fencing ready to operate the system emergency stop. A formal process of transferring responsibility to these two workers would be of aid in emphasizing the hazardous steps being taken in such an operation. A permit to Work System is envisaged as a means of achieving this. It should be noted that the above ETA does not yield any information on the working area guarding because this part of the cell is not involved in this event tree. Therefore a hazard analysis of the cell would involve considering several initating events to examine all activities and safety aspects of the cell. Furthermore, an adequate safety system for the cell would require an iterative application of ETA. First ETA could be used to identify the hazards of the cell graded in importance by quantitively evaluation [1, 2] the event tree. ETA may then be used to see if the compensating safety features are actually sufficient to prevent the hazards from being encountered.

For completeness it is worth mentioning that the working area may be protected by a light guard, which is inactive during loading operations but active at all other times.

5. CONCLUDING REMARKS

In this paper we have attempted to systematise the application of reliability methodologies to robot safety with a view to developing standards. Fault tree analysis can be used to (i) quantitatively assess reliability and thus help to define stands for all hardware (ii) identify critical areas needing particular attention, which may be tackled by defining standards or used for the preparation of checklists. (iii) evaluate quantitatively the interaction of the robot with other machinery with the limitation that time dependencies are not included. (iv) ascertain in a logical but descriptive manner the possible effects of human actions. Event tree analysis can be used to (1) assess the hazards that may be encountered in a robot cell (ii) assess the efficacy of compensating safety features of the system.

The disadvantages of both FTA and ETA are that (i) they both lead to very large trees, (ii) the choice of events in ETA and that of the top event in

FTA are crucial as careless choices can lead to obscuring safety issues rather than illuminating them (iii) both require very detailed knowledge of the system and finally (iv) good quantitative data on failure probabilities [5] is needed to produce sound figures for the hazards identified.

ACKNOWLEDGEMENTS
We would like to thank Professor Tom Husband and members of the National Centre of Systems Reliability for their interest in and assistance with this work. FD would like to thank the Science and Engineering Research Council (UK) for their support for this work and also for an Advanced Fellowship and KK for a Research Studentship. RHJ thanks the SERC and the Social Sciences Research Concil Joint Committee for a Research Studentship.

REFERENCES
1) Green, A. E. and Bourne, A. J.   Reliability Technology, John Wiley, 1972.
2) MacCormick, N. J. Reliability and Risk Analysis: Methods and Nuclear Power Applications. Academic Pres, 1981.
3) Khodabandehloo, K, Duggan, F, Husband, T. M., Reliability of Industrial Robots: A safety viewpoint. 7th British Robot Association Conference, May 1984, Cambridge.(UK)
4) Jones, R. H and Khodabandehloo, K.   The performance of robots in industry.   Young Operational Researchers Conference, April 1984, Nottinham and to be published.
5) The National Centre of Systems Reliability Data Bank, 1984, UKAEA, Safety and Reliability Directorate, Wigshaw Lane, Culeheth, Warrington, WA3 4NE.

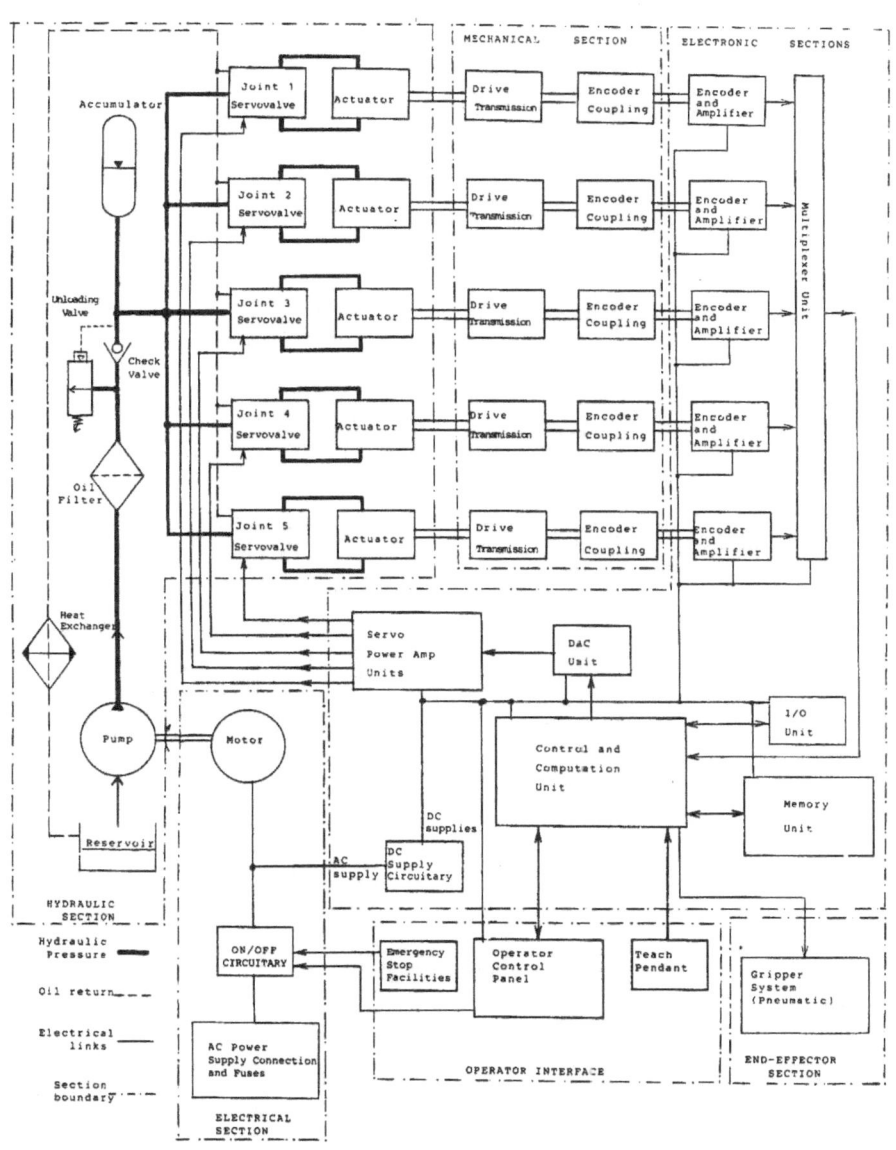

FIG. 1   SIMPLIFIED BLOCK DIAGRAM FOR AN HYDRAULIC ROBOT

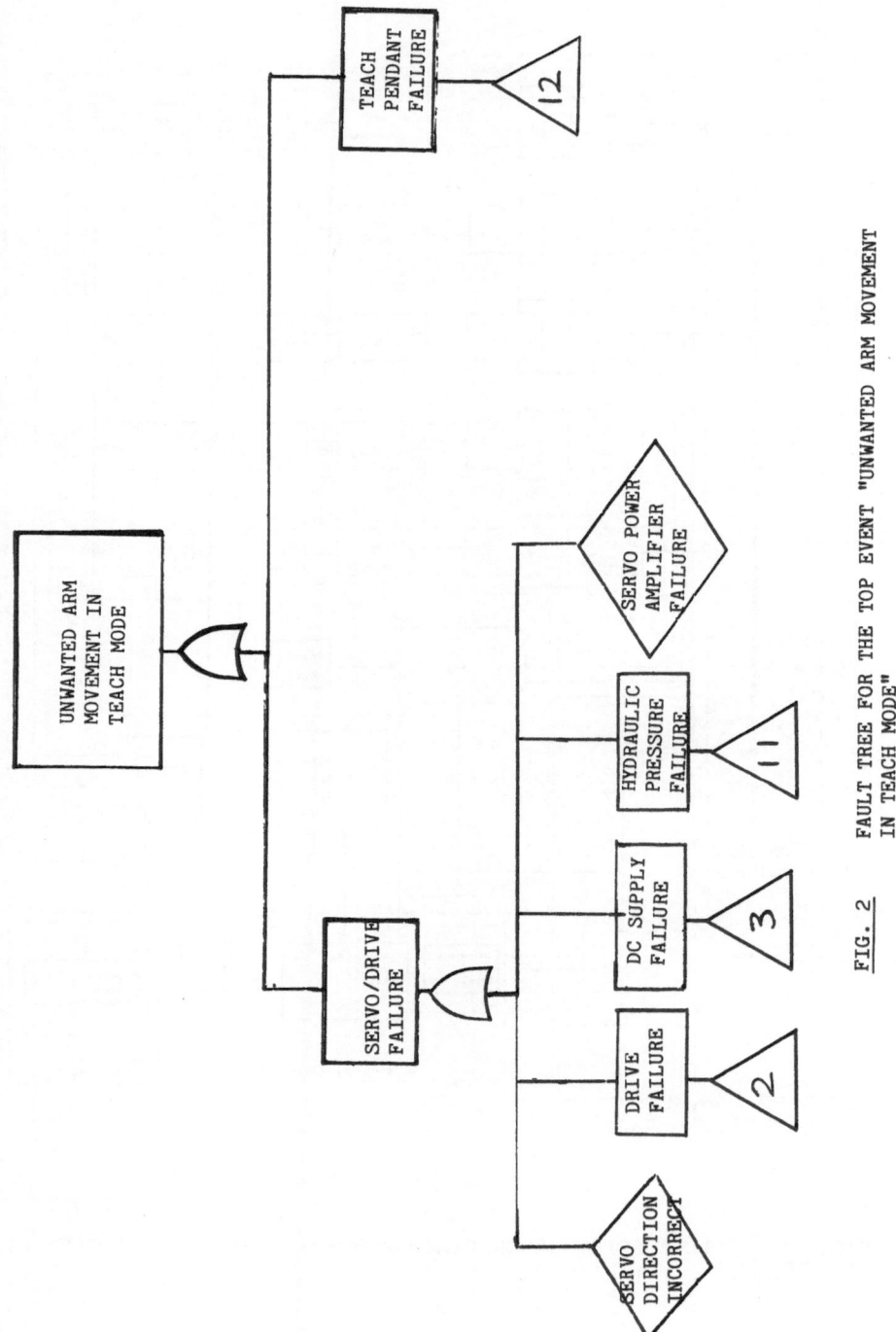

FIG. 2 FAULT TREE FOR THE TOP EVENT "UNWANTED ARM MOVEMENT IN TEACH MODE"

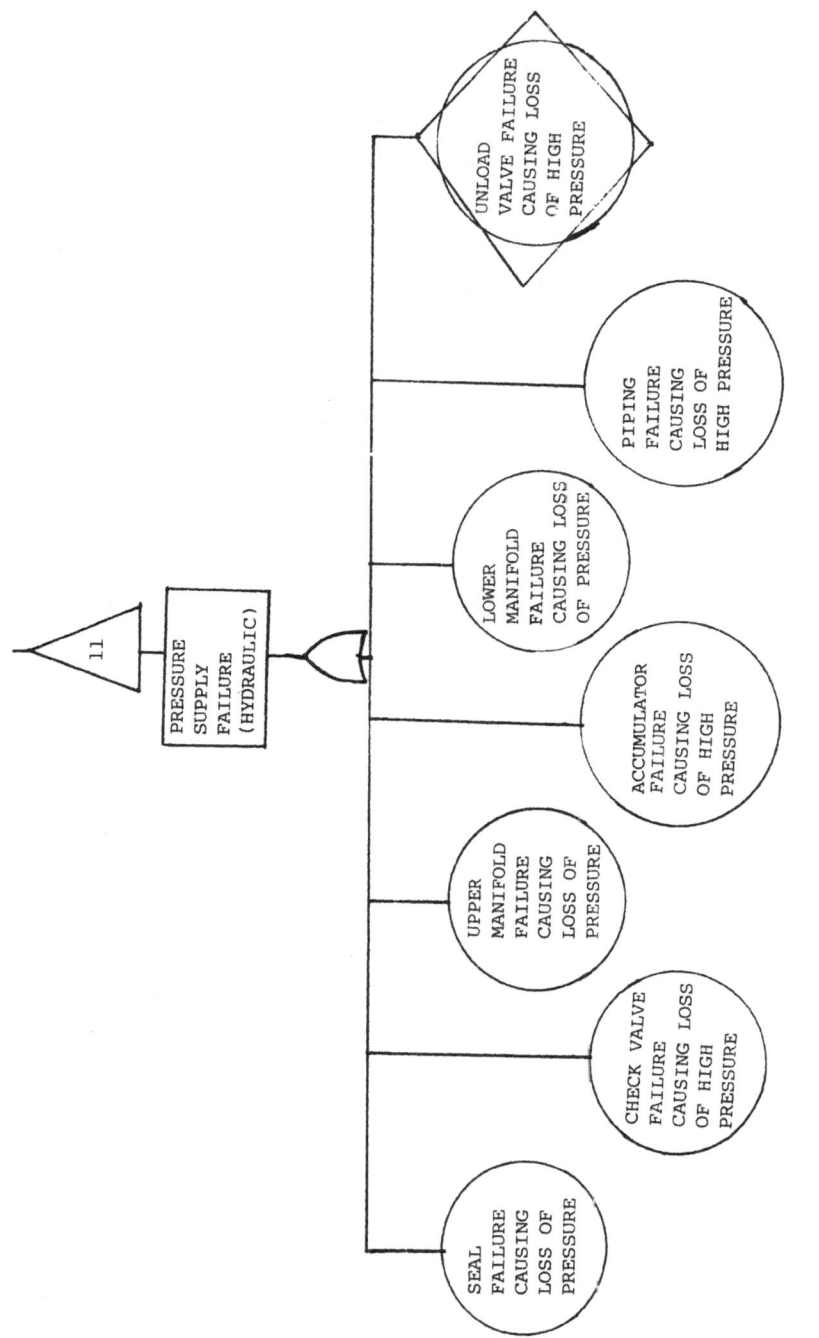

FIG. 3   SUBTREE OF FIG. 2

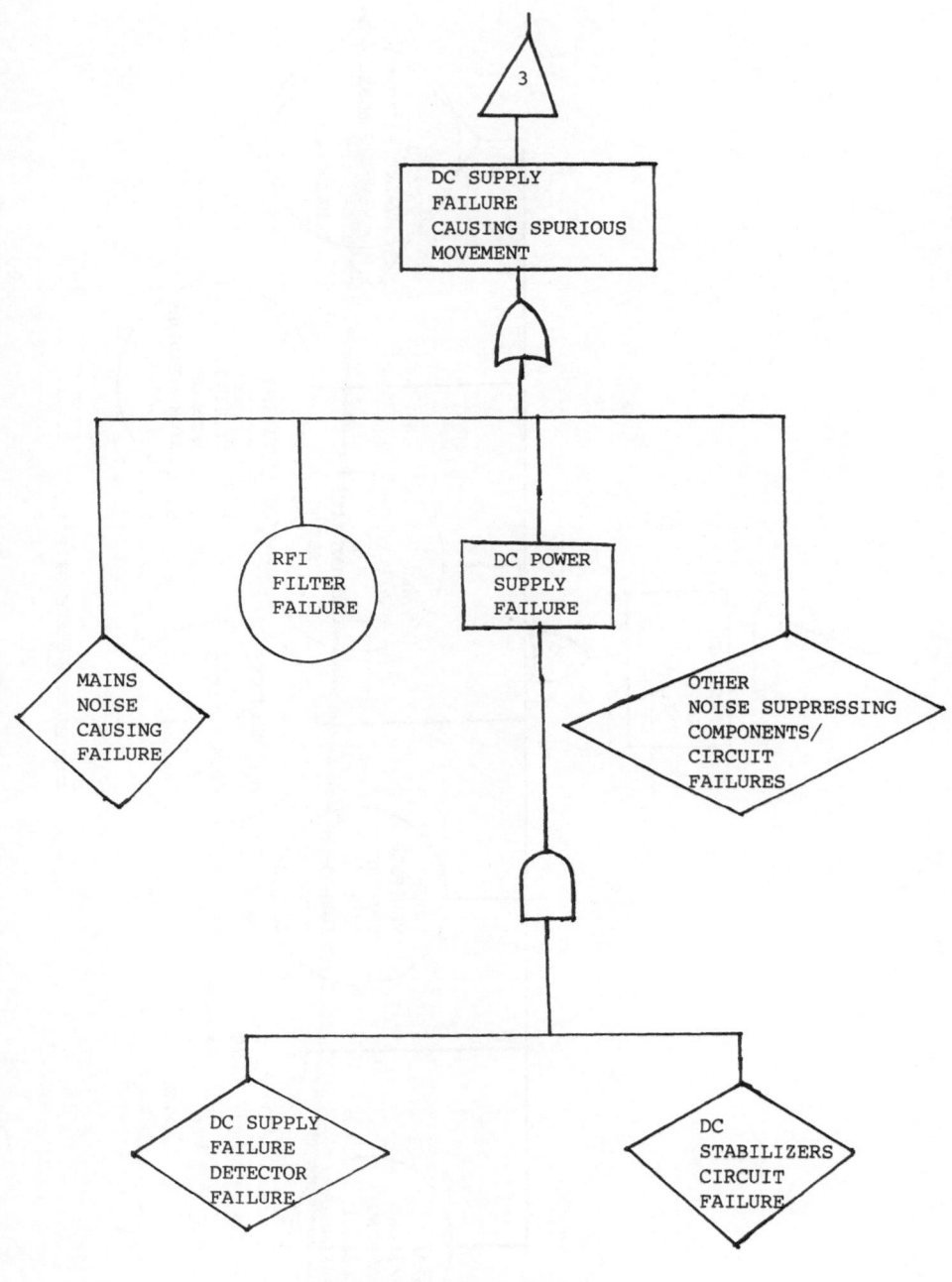

FIG. 4   SUBTREE OF FIG. 2

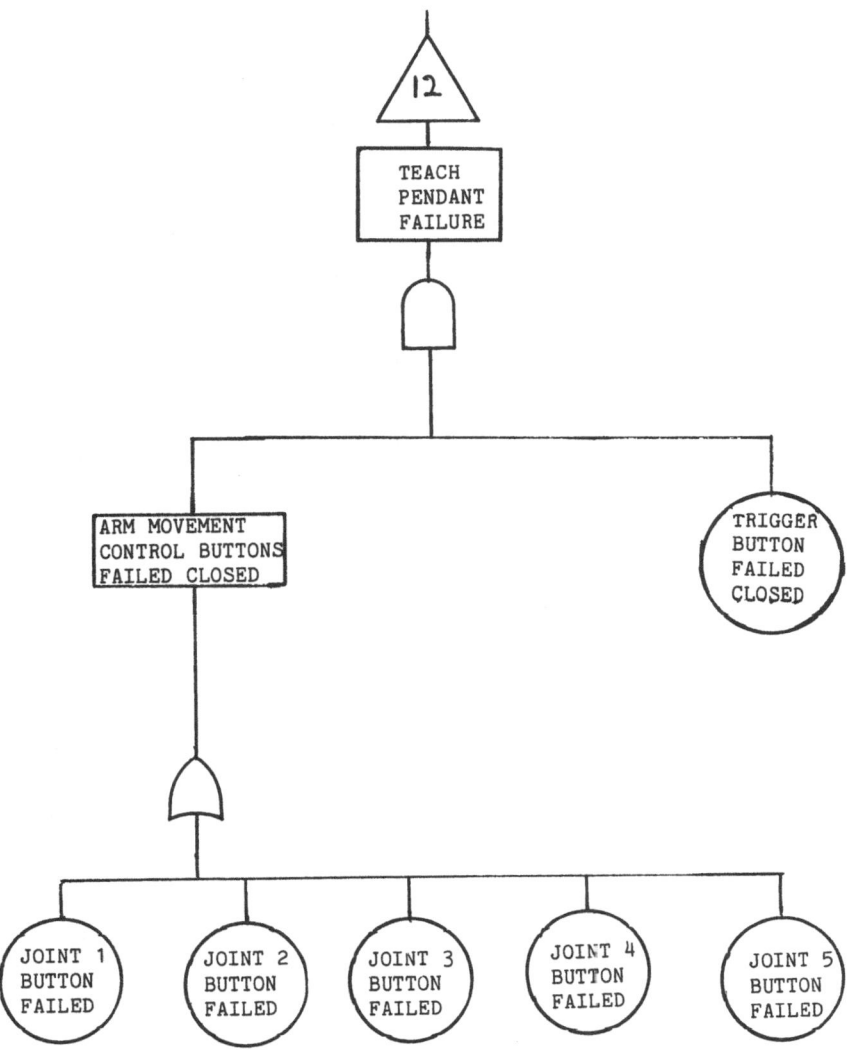

FIG. 5   SUBTREE OF FIG. 2

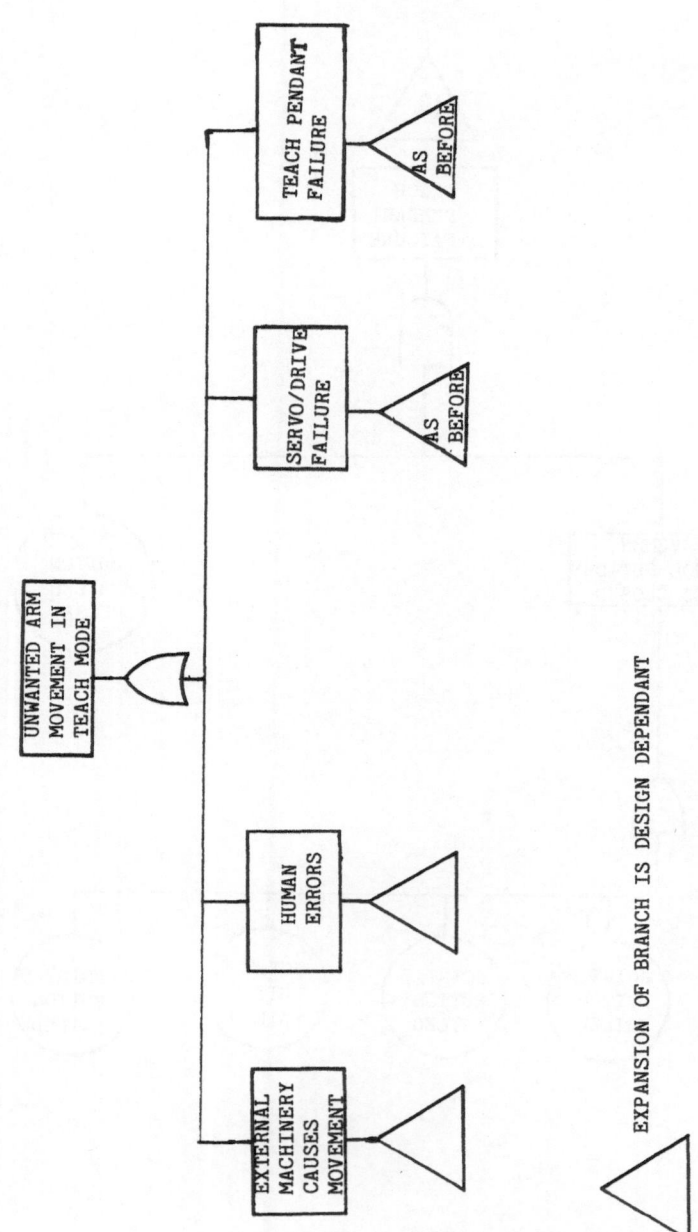

△ EXPANSION OF BRANCH IS DESIGN DEPENDANT

FIG. 6   FAULT TREE OF FIG. 2 EXTENDED TO INCLUDE EFFECTS OF
OTHER MACHINERY AND HUMAN ERROR.

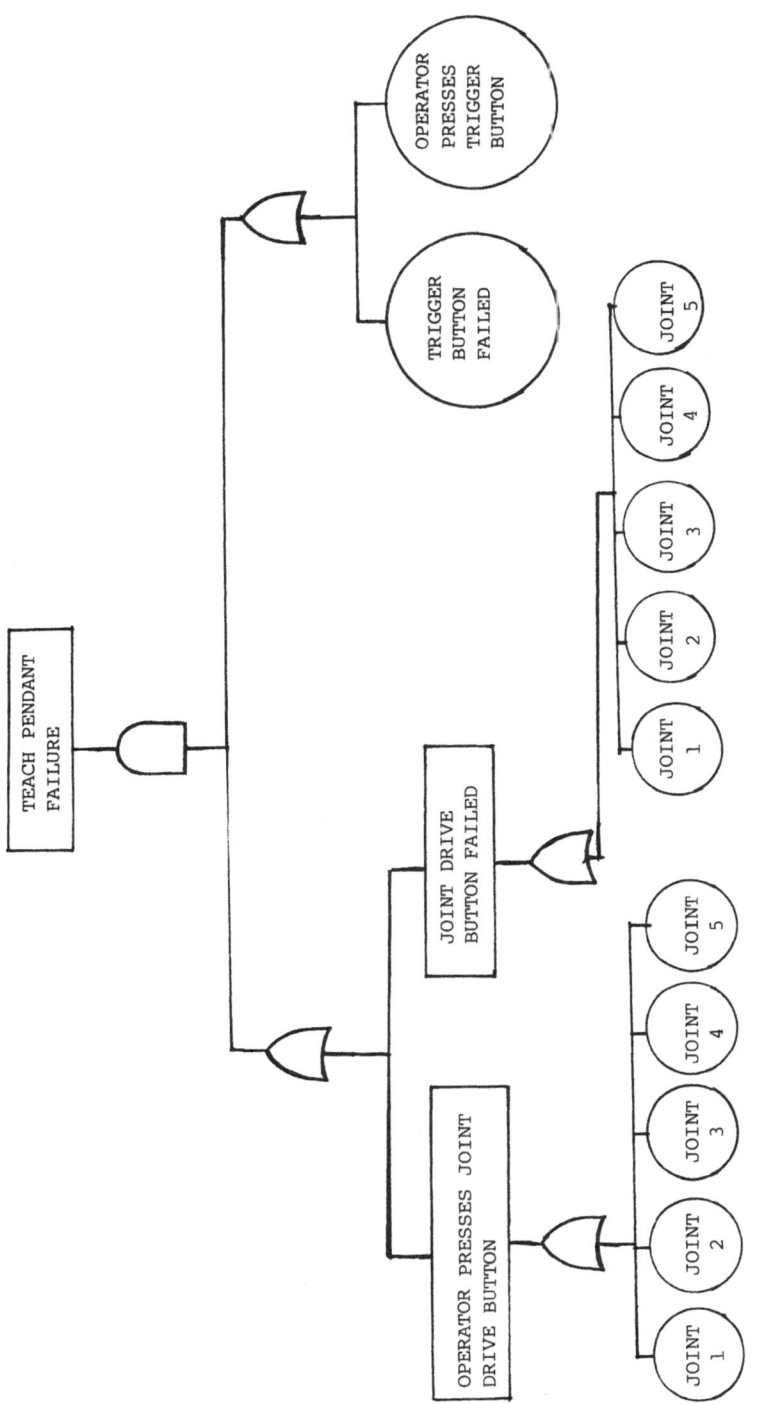

Fig. 7   FAULT THREE FOR TEACH PENDANT FAILURE INCLUDING HUMAN ERRORS.

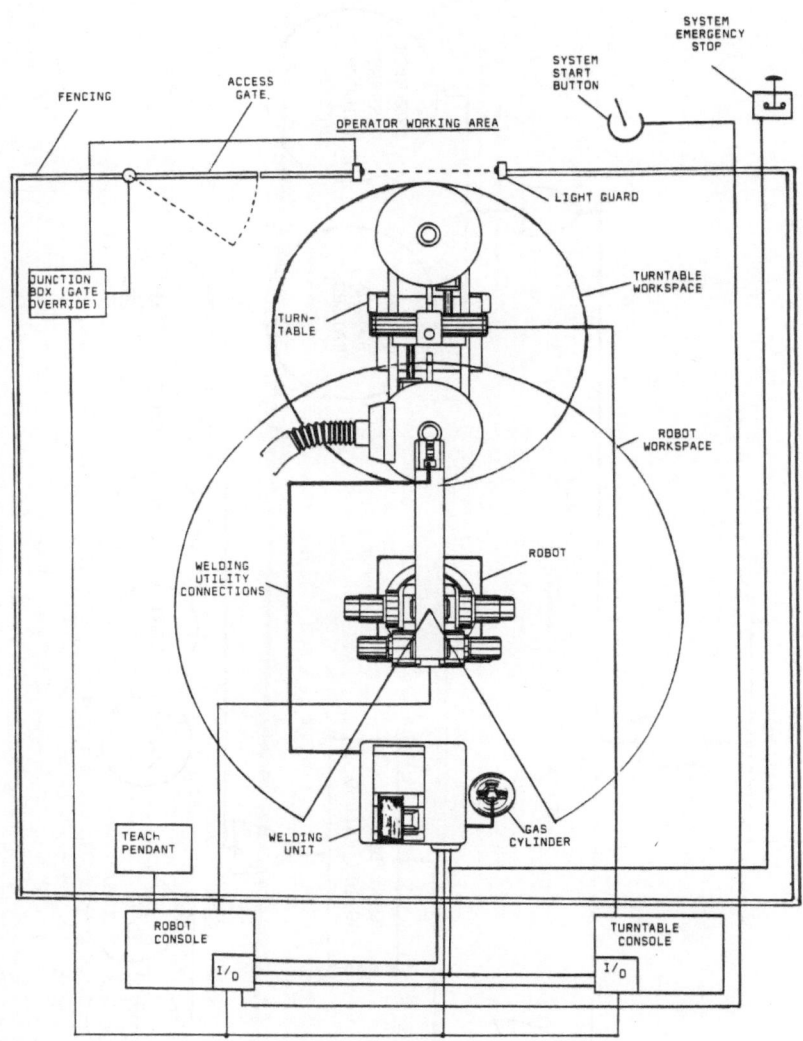

FIG. 8   ROBOT WELDING SYSTEM LAYOUT (SCHEMATIC)

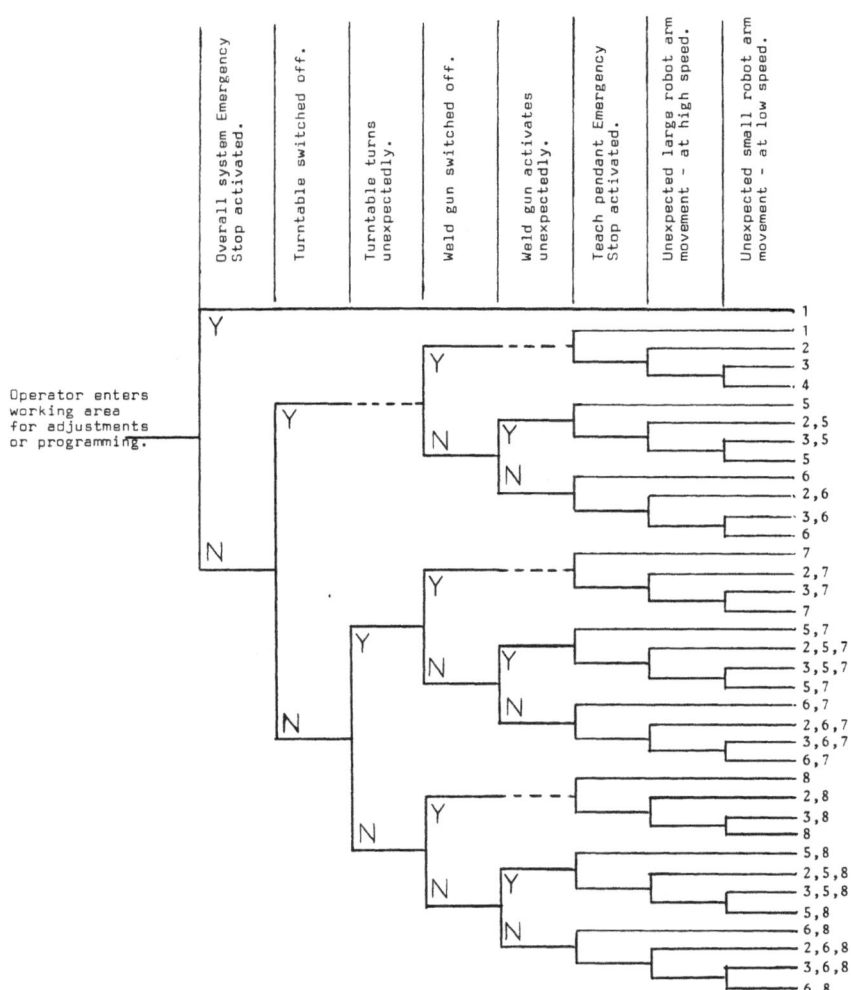

Column headers (top to bottom):
- Overall system Emergency Stop activated.
- Turntable switched off.
- Turntable turns unexpectedly.
- Weld gun switched off.
- Weld gun activates unexpectedly.
- Teach pendant Emergency Stop activated.
- Unexpected large robot arm movement - at high speed.
- Unexpected small robot arm movement - at low speed.

Operator enters working area for adjustments or programming.

1) System inactive/put into safe mode.

2) High level of hazard due to robot collision with equipment or operator.

3) Hazard due to robot causing equipment damage and possible operator harm.

4) Task performed satisfactorily.

5) Hazard from weld gun (e.g. weld flash, high voltage).

6) Potential hazard from weld gun - 'lucky escape'.

7) Hazard from turntable motion.

8) Potential hazard from turntable motion - 'lucky escape'.

FIG. 9   EVENT TREE FOR INITIATING EVENT OF THE OPERATOR ENTERING WORKING AREA FOR ADJUSTMENTS OR PROGRAMMING.

International Standards Activities in the Field of Industrial Robots

J J HUNTER

National Engineering Laboratory
East Kilbride
Glasgow
Scotland
United Kingdom

Summary

Work on the development of international standards for industrial robots
was begun in 1979. Details are given of progress to date and the enlarged
work programme to be tackled under the recently formed ISO Committee TC184
- Industrial Automation Systems.

Introduction

International standards work in the field of robotics began at the meeting
of the ISO Committee TC97/SC8 held in Berlin in October 1979 when delegates
from France, Germany, Switzerland, USA and the UK produced a workplan[1]
for the development of industrial robot standards.

The reason for undertaking this work in TC97/SC8 was that this was the sub-
committee concerned with standards for the numerical control of machine
tools. It was felt that since the structure of industrial robots was
similar to numerically-controlled machine tools and since it was foreseen
that robots would increasingly be integrated with machine tools in production cells and automated factories it was appropriate that their standards
development should be closely related.

Following this inaugural meeting, industrial robot standards were the
subject of several meetings of a working group (WG2) of SC8 resulting in a
draft standard[2] (DP8375) which has been issued for comment and approval
by the member bodies of ISO.

With rapid develoments now occurring in factory automation, SC8 and an
associated sub-committee SC9 (Numerical Control Languages) were, in
December 1983, upgraded to form the basis of a new committee TC184 -

Industrial Automation Systems. Within TC184, industrial robot standards have been promoted to full sub-committee status (SC2).

At the first meeting of SC2, at Frankfurt in May 1984, work will continue on the draft standard[2].

Workplan - Industrial Robots

In the workplan[1], industrial robots were defined and this was followed by a description of the scope of the work and an outline of how it should be carried out.

The scope of the work was taken to be standardisation of definitions, structure, operation and programming. The workplan suggested this be tackled in the order, terminology, classification, performance including safety and interface standards.

With regard to terminology, it was intended that the work undertaken should be to extend the glossary of terms[3] already established for NC machine tools.

The Draft Standard - DP8373

Much confusion has resulted from the lack of a clear definition of industrial robots. This has been shown for example in widely differing figures of industrial robot production that have appeared. The definition in DP8373 is given below. It is to be expected that this definition will be further amended in view of comments from ISO member bodies.

The other main sections in DP8373 are characterisation and graphic representation; these are summarised with comments in following sections.

"Definition of an industrial robot

An automatic servo-controlled reprogrammable multifunctional manipulator having multiple axes, capable of handling materials, parts, tools or specialised devices through variable programmed operations for the performance of a variety of tasks.

Comment 1 – The industrial robot often has the appearance of one or more arms ending in a wrist; its control unit uses a memorising device and sometimes it can use sensing and adaptation appliances that take account of the environment and circumstances. These multipurpose machines are generally designed to carry out repetitive functions and can be adapted to other functions without permanent alterations of the equipment.

Comment 2 – Figure 1 shows how industrial robots are related to other motion devices."

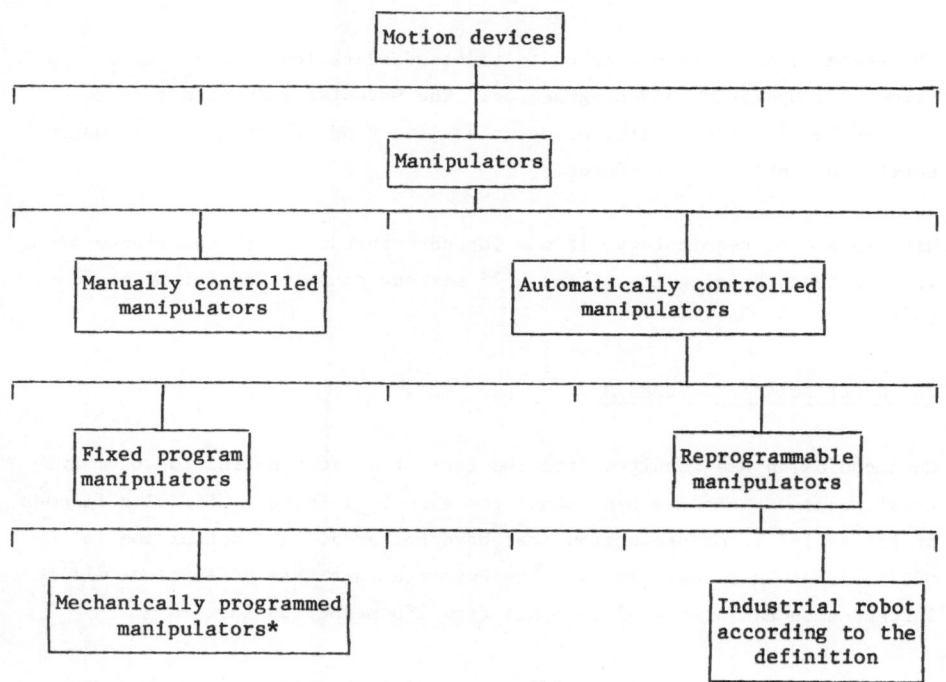

*previously often called pick-and-place robots

FIGURE 1 – CHART GIVING THE BOUNDARY FOR INDUSTRIAL ROBOTS

The definition attempts to distinguish industrial robots from other numerically-controlled machines and also from more elementary manipulators. For example an industrial robot has a control system which is similar to

that of a numerically controlled machining centre but the robot differs in
its multi-functional aspect. That is, the same robot might be used for
machining type tasks or for arc welding or for assembly tasks.

There is a close resemblance in mechanical structure between industrial
robots and mechanically-programmed manipulators (sometimes called pick and
place robots). Despite this, the difference is great since generally the
control system of the mechanically-programmed manipulator is much simpler
thus making it incapable of the variety of tasks which can be performed by
an industrial robot. Nor is the mechanically-programmed manipulator
'reprogrammable' in the sense of the industrial robot which can have path
motion data loaded into its memory from an electrical signal input port.
This enables the industrial robot to be used for small batch production as
well as for mass production.

Characterisation

It is proposed that industrial robots can be characterised by the following
attributes:

Main power source: eg pneumatic, hydraulic or electrically powered
actuators.

Type of control: point-to-point, continuous path control.

Programming methods: teach programming; off-line programming.

Types of interaction with external processes: branch interaction where
external switches can select different program sequences; feedback interaction where the robot path can be modified by sensors; adaptive interaction where the external process can alter the robot control system parameters.

Kinematics: the configuration of the joints and linkages of an industrial
robot determines the path through which it is possible to move its end-
effector.

Performance: working space; load capacity; accuracy.

Agreement on what constitutes important characteristics allows robot information to be presented in a concise and consistent manner. The characteristics given in DP8373 are general and capable of being extended for future requirements.

Graphic representation

A set of symbols for representing the kinematic configuration of an industrial robot is proposed (Table 1). It is probable that these symbols will be further revised.

Future Work Programme

At the inaugural meeting of TC184, an ad hoc group, with members from many countries proposed the setting up of an 'Industrial Robots' Sub-committee SC2 of TC184 which would continue and extend the work of WG2 of TC97/SC8. The areas of work to be covered would be:

- Definition - characterisation.

- Terminology.

- Graphic representation.

- Performance and performance testing methods.

- Safety.

- Mechanical interfaces.

- Programming methods.

- Requirements for information exchange.

The secretariat of SC2 will be held by AFNOR

To speed up the work of SC2, a number of working groups were set up. These together with their convenors are:

WG1 France   - Definition, characterisation, terminology graphic
representation.

WG2 Sweden   - Perforamnce and performance testing methods.

WG3 Japan   - Safety.

WG4 USA   - Mechanical interfaces.

WG5 France   - Programming methods.

WG6 Germany - Requirements for information exchange.

Programming methods (WG5) encompass all aspects cf programming including teach programming and off-line programming using explicit or goal directed programming languages. Non device-specific language aspects will require to be co-ordinated with the work of the Application Languages Sub-committee SC3 of TC184.

Requirements for information exchange (WG6) will similarly encompass the whole area of man - machine, and machine - machine communication in relation to industrial robots. WG6 is expected to liaise with the Communications and Interfaces Working Group GT1 of TC184 (note: GT1 became a working group rather than a sub-committee because no member body would undertake the secretariat). Much work has already been done on communications and interfaces in ISO, IEC and ITU. The work of GT1 is therefore likely to be based on prior work in Committees ISO TC97/SC6 and IEC/TC65.

A number of technical documents have been submitted and await consideration by SC2. These include proposals on:

Terminology (France, USA, Japan).

Axis notation, geometry, graphic representation (France, Germany, Sweden, Japan).

Performance testing (France, Germany).

Standard form for indicating characteristics and functions of industrial robots (Japan).

Safety (Germany, Japan, UK).

In their report following the inaugural meeting, the ad hoc group also requested further contributions on:

● programming methods,

● requirements for information exchange, and

● mechanical interfaces.

Conclusion

The structure of the new ISO committee TC184 - Industrial Automation Systems, affords a favourable environment for the development of industrial robot standards that are related to those of the other machines and equipment with which they will have to operate. Although much work remains to be done, a sound framework has been devised under which this development can proceed.

Acknowledgements

This paper is published by permission of the Director, National Engineering Laboratory, East Kilbride, UK. It is British Crown copyright.

References

1   ISO/TC97/SC8-451, Workplan - Industrial Robots, November 1979, ISO Secretariat, AFNOR Tour Europe, Cadex 7, 92080 Paris La Defence.

2   ISO-DP8373, Industrial Robot - Definition, Classification and Graphic Representation, July 1983, ISO Secretariat - AFNOR, Tour Europe, Cedix 7, 92080 Paris La Defence.

3   ISO2806-1980, Numerical Control of Machines - vocabulary.

DP 8373

Table 1 - SYMBOLS USED TO REPRESENT THE MECHANICAL STRUCTURE OF AN INDUSTRIAL ROBOT

	FRONT VIEW	SIDE VIEW	TOP VIEW	ISOMETRIC VIEW
PRISMATIC JOINT TELESCOPIC				
PRISMATIC JOINT TRANSVERSE				
ROTARY JOINT PIVOT				
ROTARY JOINT HINGE (a)				
ROTARY JOINT HINGE (b)				
Example of a DISTRIBUTED JOINT				
COUPLING DEVICE				
BASE				
GROUND				
BASE WITH GROUND				
Example of a CONCURRENT AXES subassembly				

Standardization in the Industrial Robot Field

B. KNOERR

VDMA (the German Machinery Manufacturers and
Plantmakers Association)
Frankfurt, Germany F.R.

Introduction

Standardization and industrial robots technology - at first
sight relatively contrary activities: on the one hand the
conservative image of standardizing only proved technical
practice and on the other hand the sophistication and inno-
vation of robot systems and related technologies as information
processing in manufacturing systems. Why has the development of industrial robots standards become so much attraction?

To answer this question some important facts should be illumi-
nated.

Why Standardization?

In principle standardization is nothing else than the realization, that our today's world (and especially our technical
evironment) needs interpretation facilities assuring the prevention of misunderstanding or misinformation and to avoid
very expensive defects and, even more importantly, safety risks.

For better understanding the following remarks are pertinent:

- Economic aspects: The world has become small and smaller with
  the result of intensive trade exchange and trade interlockings
  under different conditions of competition. In this scene
  the lack of at least basic international standards would create
  confusion between users and manufacturers (in respect of comparison of commercial conditions, technical specifications
  etc.) whilst the existence of only national standards will
  produce   barriers to trade (whether intended or not).

On the other hand an abundance of rules can be a cause for an "equalization" of competition impacts.

Another important influence may come from the progressive tendency to raise technical standards to a status of legal regulations with all consequences regarding product liability and supply contracts conditions.

Standardization is reducing problems of product interchange-ability by unifying mechanical and/or physical interfaces, it generates larger product volumes by means of (mostly) reduced prices and supports and in this way, the acceleration of market implementation.

- Technical aspects: On the basis of technical performance data, experts have to decide whether or not equipment may be suitable for relevant tasks. Basic needs for this procedure are the existence of an exact terminology, the detailed definition of terms and qualified methods of determination of characteristics. These requirements can only be accomplished by means of quali-fied standards, which may protect the potential user from un-realistic expectation of performance data and disappointing results of the finished installation.

- Safety aspects: When talking about new production technologies, safety and human factors are of growing interest. It is an important standardization task to transmit the results of safety research and practical experience into rules suitable to reduce safety risks as far as possible.

Who is developing Standards?

Basically national organizations are responsible for developing standards being valid in their region of activity. Some examples: AFNOR in France, BSI in Great Britain, DIN in Germany F.R. etc.

Nearly all these national organizations have agreed to support and accept the international standardization work which is conducted by the International Organization for Standardization ISO (resp. IEC for the electrical part of standardization).

Does Standardization Support or Prevent Progress?

Some facts seem to be unavoidable: The standardization work is guided by a frame of formalism such as national and international rules and directives as well as pronounced national interests and even personal ambitions and "self-presentation". This initiates streams which are not qualified to support an effective and welltimed standardization basis which is welcome by both users and manufacturers.

National or European normalization solos, the production of too many restricting standards and the mistiming of start and finalization of work are factors which prevent technical and commercial progress especially in new technologies. However, a welltimed and organized ISO-standardization work corresponding to the industrial needs would certainly support technical progress and the extension of applicability of technology.

From these more general points of view we return to the standardization in the industrial robots field.

Organization of the Robotics Standardization Work

Beside the various national efforts some international initiatives have been taken in the late 1970's. Mainly the French ISO member body started the discussion within the Numerical Control Systems Committee (ISO/TC 97/SC 8) by presenting a work plan taking necessary tasks into consideration.

Subsequently a Working Group within the before mentioned Sub-Committee has been founded and supported by experts of important industrial countries such as:
France, Germany F.R., Great Britain, Hungary, Japan, Norway, Sweden, Switzerland, USA.

Meetings have taken place in Paris, Zürich, Stockholm and Frankfurt. In December 1983 a new ISO-Committee TC 184 "Industrial Automation" has been established, which will be the parent organization for robotic standardization activities in the near future. For May 1984 the constitution of ISO/TC 184/SC 2 "Industrial Robots" is planned, which will be divided in various working groups with responsibilities for the standardization of definitions, terms, characteristics, graphic representation, safety requirements, programming methods (in coordination with further TC 184 - Sub-Committees) and mechanical interfaces in the field of industrial robots and related handling equipment.

The industrial robots committee will be working under the chairmanship of Monsieur Jean Chabrol, France, and with the Secretariat of AFNOR.

Update on Robotics Standards Main Tasks

This chapter should be devided in two parts:

1. Stand-alone hardware:
   To support the understanding and implementation of robotic technology the treatment of basic tasks is urgently needed, containing the fields of terminology, the definitions, graphic representation, performance data incl. test methods, safety requirements and programming languages.

2. Integration of automation technologies:
   Very significant is the trend away from stand-alone units to production plants with a high degree of intelligence and integration into their environment and surrounding equipment. This trend requires intensive standardization work in connection with high-level information processing and data communication as well as the clarification of mechanical and electric/electronic interface conditions and, last not least, criteria for the human situation in highly intelligent automation production plants.

Statements on the current Robotic Standards Situation

Certainly typical of the current optimistic atmosphere in view of the international robotic standards development are the following statements:

Jean Chabrol, Renault, Paris and Chairman of ISO/TC 184/SC 2 "Industrial Robots":
"1. International Standardization is a must, especially when dealing with a new equipment such as industrial robots, for the understanding and clarification of new technical situations.

2. By giving guidelines the international standardization helps improve new equipment which will have to be compatible and in our future industrial world this consideration is of paramount importance.

3. International Standardization should be worked out of a consensus of all participants: It's a link between persons involved in this technique without consideration of competition or trade."

Peter Brown, Deputy Director, Standards Development Division of the American Society for Testing and Materials (ASTM), in an interview with "Robotics Today", October 1983: "Our committee is tracking the activities of ISO on robotics closely. There is a French and a German document on terminology and performance tests. If some good work has already been done under the ISO banner, perhaps certain features should be incorporated in ASTM-standards. The strongest position for the U.S. is to be aware of standards development worldwide and to interact on a global basis."

The author's position: In Germany all groups interested in the field of robotics standardization, e.g. DIN (the Standardization Organization), VDI (the Association of German Engineers), VDMA/MHI (the Association of German Robotics' and Handling Equipment Manufacturers), the Trade Employers' Associations etc., are "placed around one table" to ensure effectiveness of work and to avoid parallel or opposite activities and targets. Unanimously the opinion is dominating that no national standards projects are justifiable if the ISO standardization work runs well and effort should be concentrated to support this activity.

Conclusion

When talking about the implementation of automation technologies such as industrial robotics, all participants (users, manufacturers, scientists etc.) should be in the same boat with the following aims:
- to recognize that standardization is necessary
- to achieve progress of work
- to concentrate the experts' capacity upon the international

standardization and to avoid national (and also European) attempts

- to standardize only as few as necessary and as much as justifiable.

International Standardisation Related to Industrial Robots

H. TIPTON

Machine Tool Industry Research Association,
Macclesfield

Summary

The current activities of the IEC and ISO in the field of standardisation
of industrial robots are reviewed.

Introduction

The robotics industry is, like that of numerically controlled machines, an
international one dominated by a few large companies but with considerable
activity in most industrial countries.

Robots are used as components of a system and there is a need for
compatibility with other parts of the system which can only be met by a
large degree of standardisation. With robots this is desirable in many
aspects - data formats, interconnections, safety, performance, etc. so the
task of standardisation is not a small one.

The two bodies mainly concerned with standardisation in industrial
automation are the International Standards Organisation (ISO) and the
International Electrotechnical Commission (IEC).

ISO

ISO work on standardisation of industrial robotics has evolved side-by-
side with that on numerical control, within Technical Committee 97 -
Computers and Information processing, Sub-committee 8 - Numerical control
of machines and Sub-committee 9 - Programming languages for numerical
control.

Submissions on industrial robots were received from 1978 onwards and led
to the formation of a working group ISO/TC97/SC8/WG2 on robots under the
guidance of Mr. Chabrol.

The evolution of industrial automation and in particular of flexible manufacturing systems incorporating control computers, numerically controlled machine tools, robots, inspection machines etc. meant that the scope of TC97/SC8 and SC9 together did not cover adequately the standardisation requirements of industrial automation.

This led to the formation in 1983 of a new Technical Committee 184 Information processing systems as related to industrial automation/ numerical control of machines with five sub-committees and one working group responsible to the technical committee:

ISO/TC184/SC1 Numerical control of machines

/SC2 Industrial robots

/SC3 Programming languages

/WG1 Communications and interconnections

/SC4 Exeternal representation of product definition data

/SC5 Requirements for system integration

These sub-committees ensure the continuation of the work of TC97/SC8 and TC97/SC9 and that the current information processing requirements of industrial automation are adequately covered.

At its first meeting in Frankfurt on 22/24 May 1984 ISO/TC184/SC2 established four working groups:

WG1 Terminology and graphic representation

WG2 Performance criteria and related test methods

WG3 Safety

WG4 Programming methods and data communications

IEC

The main distinction between ISO and IEC is that the latter concentrates on standards in the electrical and electronic fields while ISO covers a wider range of subjects.

In the field of numerical control the standards relating to data formats and coding of information have been developed within TC97 of ISO but the standard on the "Interface between numerical controls and industrial machines" is IEC 550 (1977) and it is concerned with methods of inter-connection, voltage levels etc. to ensure safety.

The relevant IEC technical committees are:

TC44 Electrical equipment of industrial machines,   and

    TC65   Industrial process measurement and control,
with sub-committees:

    SC65A System consideration

    SC65B Elements of systems

    SC65C Digital data communication for industrial process measurement and
           control systems

In the current (1983) publications of the IEC there is no mention of robots
as such but it is understood that working group 1 of IEC TC44 has proposed
that TC44 should

"prepare a draft standard which will provide a detailed specification
or give guidance relating to the electrical/electronic equipment of
industrial machines and industrial robots controlled by PES's
(programmable electronic systems)."

Cooperation

There has always been a need for contact and cooperation between ISO and
IEC, and indeed between the various sub-committees of both organisations
but with the increasing use of information processing the boundaries of
work of the two organisations are not as clear-cut as they used to be.

Contacts had previously taken place by joint membership by individuals of
committees and by the attendance of observers at each others meetings, but
the formation of ISO/TC184 on industrial automation has brought this out
into the open.

The inaugural meeting of TC184 in Paris at which its scope and programme
of work was decided, was attended by representatives of the Secretariat of
the IEC and in Frankfurt on 22 May 1984 during the meetings of ISO/TC184/
SC1 and SC2 a discussion was held between the IEC and the ISO on the
programme of work of TC184. It was decided that there was no overlap of
work at the moment and a number of areas in which there was a need for
liaison were identified and measures were taken to establish the liaisons
including the exchange of documents and attendance at each others meetings.

Conclusion

The two international standards bodies, the IEC and ISO can be seen to be
responding effectively to the needs for standardisation in the field of
industrial automation.

This is a particularly difficult area of work because of the high rate of change of modern technology and the relative slowness of the generation and production of standards, but against this must be balanced the reward of sound international standards in lubricating the flow of trade and progress.

Robot Languages in the Eighties

Giuseppina Gini
Dept. of Electronics, Politecnico, Milan, Italy

Maria Gini
Dept. of Computer Science, University of Minnesota, Minneapolis, USA

ABSTRACT

The scenario of general-purpose programming systems is rapidly changing; what are the consequences for robot programming? The programming environments built around ADA, UNIX, and Interlisp are useful for robot programming? After introducing the peculiar aspects of robot programming we will discuss some examples of general-purpose languages applied to robots and languages specifically designed for robotics. Criteria for making a choice between the two approaches should take into account the present state of the art. The need of a strong integration between different components as robots, vision systems, and other automation equipment could support the first approach. The solution of robot and robot users problems has, until now, supported the second approach, as indicated by the choices of European robot manufacturers. Anyway the expression of actions taken by different intelligent agents, as robots can be defined in the future, will require absolutely new linguistic media. We hope that robot programming in the eighties can develop experiences useful for the assessment in the field.

INTRODUCTION

Today robot applications are generally carried out in an integrated industrial setting, where robots and other equipment manipulate and sense parts to repeatedly perform a task. We may see a typical setting as consisting of a conveyor belt which transports parts, usually partially oriented and separated, a vision system getting information about the incoming parts, and a robot making assembly or inspection or serving other machines. Those robots are sometimes programmed on line, guiding them through their task and storing positions and operations into a memory. The drawbacks of this methods have been stressed enough. Errors during the programming require to restart again, teaching of many positions is too long and error prone, the programming time is not productive because all the robot related equipment must be stopped during new task programming, synchronization with other equipment is hard, sensors cannot be used to modify actions during program execution. During recent years we have seen a significant change in the attitude of robot manufacturers, and almost every new robot is today sold with an off line programming system. Ten years of experiences have passed and manipulator level languages are today accepted. Usually we use them to give the cartesian reference frames where the manipulator hand should be moved in order to accomplish the task. Those frames or their sequence can be modified by run-time events, as sensor output or external synchronization signals. In some early systems every joint of the manipulator was individually given a value in its own coordinate system. Programming in terms of frames is not yet a good solution because it requires a lot of detailed information be given by the programmer, who is expected to have some skills in mathematics and programming. The writing of robot programs is not so easy as one might think. It is quite difficult to understand and use positions in 3D space, and this is the basis of robot programming now.

An intelligent robot (20) will be given only a task and the spared parts and materials needed to perform it, instead of programs embedding the complete sequence of actions to be taken. An issue to be addressed by more advanced systems is world representation and object modelling. A representation of the world should allow the robot to manipulate parts and to sense the environment. This model should be a copy of the real world. Actually no world model can be complete, so at least it should be detailed and rich enough. Present industrial robots do not have any geometrical world model; their knowledge is encrypted into variables and data structures which have only meaning for their human programmer. An integration of modelling used in design and production is highly desirable. Many systems are today commercially available for CAD/CAM applications. Most of them are purely graphics systems, without much interest for robotics. The models used by CAD systems may be useful in robotics, even though robot relevant information, as the center of gravity, the mass, the grasping point, etc. of the object are not contained in CAD models. The experience made in the RAPT system is that using CAD systems world modelling is still long and tedious. Moreover the use of CAD data bases for defining world models to be used also for sensorial tasks (for example for vision understanding) is not yet acceptable.

Besides these formidable research issues robots are today in use. The practical solution taken has been to limit the intelligence and the understanding of the robot and to reduce programming of robots to usual programming or, in a few cases, to parallel programming. No use of world models is done in any commercial system.

Many industrial issues in robot programming are still open. How to program robot completely off-line, and how to express the integration of different devices, being the most studied. The present status of software and software engineering has here its impact. Having removed from robot software the problem of intelligence has reduced robot software to a sophisticated software engineering problem: software solutions should work there.

ROBOT PROGRAMMING

The most obvious ways to implement robot languages are to adapt a general-purpose programming system or to develop robot specific programming systems. In the second case new languages may be developed from scratch or from existing automation languages, such as APT.

The first solution is appealing because education and training of robot programmers can be reduced in time and the development of robot applications will be reduced to the writing of a few routines. The second solution is appealing because robot programmers can be not computer expert and what they have to learn will be exactly what they need to use. In the case of APT the same NC programmers could be easily converted into robot programmers.

The first solution has as a shortcoming that it heavily relies on what is available for general purpose programming. The computer-user interface is not tailored on the specific needs. The second solution has as a shortcoming that many development efforts are often spent only to provide something already available with minor differences. In the case of APT like languages we may also argue whether robot programming is the same as NC machinery programming.

Since a language is a way to express or to test different solutions, we may say that all the developments so far obtained have demonstrated what can be done today at the manipulator level programming. After that, the choice of using existing languages or developing new ones is a problem of market image and acceptance. We may hardly see the need to develop new languages. The only important difference is in the programming environment. The development of programs off line requires sophisticated programming environments which are not generally available in standard languages as FORTRAN or even PASCAL. The choice to develop a complete programming environment is the only we see still pushing towards the development of new systems. A set of subroutines written in FORTRAN are long to edit, write, debug.

If the language should demonstrate new solutions or in other word should allow task oriented programming then we may still want to maintain manipulator level programming as target language for planning and sensorial activities. The expression of those activities not necessarily should resemble any of the existing programming languages. Graphics, natural language, or sometimes mathematical equations may be all useful in providing ways to express tasks.

Many papers have reviewed existing robot programming languages, between them (7,22). We do not intend to review all the existent robot programming systems. We want only to see what new ideas have emerged from a group of them and whether they are assessed or waiting to be fully explored.

Usually two systems are strictly integrated in a robot programming system. The *user language*, in which application programs are written, and the *run-time system* which executes the code generated by the language translator. This solution is similar, for instance, to the one used in most of the Pascal systems. It may be used as a way to standardize user languages by changing only the run-time system, as done in VAL (27), developed for PUMA robots and then implemented on other Unimation robots. Often the run time system run on micro computers and is written in assembly languages. This trend could change when more and more cheap computer power will make reasonable to write all the software in high level languages.

An issue to be newly addressed is that of defining standard software interfaces between the robot and the user level software. What kind of information we pass to the robot? joint positions, or frames? In which order? How we may ask for a point to point execution? for a continuous path? There are no reasons why a cartesian robot and a polar robot should be programmed in completely different ways. While the run time system is well tailored to the specific hardware in use, the user level language should be problem oriented more than manipulator oriented. If this standard software interface would be provided and accepted by any robot manufacturer we may get any robot language to work for any robot; homogeneity and modularity will also be valuable in industrial settings. We mention here the CAM-I project of standardization in robot software (33). They have individuated five main components of robot software: robot language, robot simulator, robot controller, robot modeler, teaching system. They have put Artificial Intelligence parts in the CAD/CAM environment, and this seems more a decision not to deal now with difficult problems as the ones open in perception and decision making. In our opinion most of them should be solved at the robot level, being the robot the flexible and adaptable entity of the FMS.

We do not intend to give here a complete list of reference terms to compare robot programming languages. What we had in mind in giving the following descriptions is based on the following key words.

```
1. expression of movements (joint level, hand level, object level);
2. expression of trajectories;
3. use of sensors;
4. class of the language ( Pascal, Basic, functional);
5. implementation (Interpreter, compiler, programming system);
6. I/O: ports, functions provided,integration with other equipment;
7. multitasking, synchronization, and parallelism;
8. integration with CAD/CAM for simulation, planning, control.
```

LANGUAGES AND SOFTWARE ENVIRONMENTS

We may look at general purpose languages as candidates for robot programming. Some experiences have been made of using Pascal for robot applications. See for instance PASRO (4) as a working example of this. Even Pascal however presents some shortcomings as a language for automation. Among them, Pascal doesn't support cooperation, which can be obtained at the operating system level, moreover, it is poor in file management, and file management would occur very often in integrated manufacturing. Other solutions have been tried, for instance using Concurrent Pascal, a small language developed around Pascal to define and execute concurrent tasks.

Languages for automation did not exist before the introduction of robots. The only exception is APT, the language for NC programming. APT in fact has been chosen as a basis for robot languages in two projects, at least, ROBEX (32) and RAPT (25). Even though, its use in robotics has not yet been demonstrated truly useful.

Software environments for general purpose programming are now available on most computers. The most complete and advanced software environments are today those of UNIX and Interlisp, while the situation of ADA is not yet assessed. The main advantage of a software environment over a simple compiler of a language is that the first provides an unified approach for all the problems encountered in the project, the development, and the test of the program. In the following we will briefly review those three systems, Unix, Interlisp, and ADA.

UNIX - The output redirection and the pipeline mechanism to connect programs are between the most useful characteristics of Unix to make modular programming easy. Different programs can be connected in any meaningful way. Unix operating system is very popular on the scientific personal workstations, and some of them are intended for CAD/CAM applications. Unix can become the standard operating system for CAD/CAM applications, perhaps not for robot programming. The computers used today to move robots usually runs both the run-time system and the user level language. To reduce the hardware costs they usually are stand alone systems, without any standard operating system. Only very sophisticated robots, not today on the market, could justify the high cost of using a sophisticated computer for their own needs.

INTERLISP - The only experience so far reported of using LISP for robot control and programming has been done at the MIT Artificial Intelligence Lab. where Mini (28), an extension to LISP to deal with real time interrupts and I/O, was used to program a robot equipped with a force sensor. LISP has not yet obtained the interest of robot manufacturers. Many reasons for that can be envisaged. Only recently LISP has been made commercially available on mini and microcomputers, it requires a lot of central memory, it is inefficient in mathematical computations and array management. Even though most of those shortcomings are true LISP can be considered now with a great interest for two reasons. The functional stile of programming which is at the basis of LISP (even though Mini is not an example of functional programming; it used a lot of SET instructions!) makes it interesting because languages embedded in LISP are completely extensible so that all the manipulation functions could be modified by the user. We will discuss the functional style of programming in the following. A second good reason for using

LISP is that robot programming tends to use more and more artificial intelligence techniques, and LISP is still considered the main Artificial Intelligence language. Serious Lisp development requires several software components still not available in even the UNIX environment. This reason makes still LISP an intensive memory user, and is the reason of developing LISP machines to make the best use of the computer. Unfortunately those machines are too expensive for any use at the factory level now.

ADA · The main motivation for using ADA (10) in robot programming is that ADA is a structured and complete computer language, and offers some advanced tools as extensibility, modularity, real time capabilities, strong type checking to increase the programs reliability. Moreover it has been designed to be the only language of the eighties and claim has been made that ADA could substitute any language from the assembler level to highest levels. The importance of a complete programming language for robotics applications has been made clear: we want robots to cooperate with other equipment, and this requires task synchronization and coordination of different and sometimes non trivial tasks. Such programming and coordination is not provided by most of the robot programming languages in industrial use today. We do not know examples of robot run time systems written in ADA while we may find examples of other subsystems developed in ADA. Vision is one of them. A long practice in vision programming has been to choose C or Pascal; ADA is a superset of them and its use should solve more problems than it can open. The experience so far developed in Ann Arbor (31) has demonstrated that the use of packages (a way to simplify program encapsulation) and generic packages (a way to implement abstract data types) can simplify the development of complex software making it easier to distribute tasks to different people, to integrate them, to modify the manufacturing cell without complex software modifications. On the other end, the big size of ADA can be a problem. Many features of ADA are hardly useful in industrial automation, but their presence makes ADA compilers big and ADA language difficult to use. In no way people able today to manually guide a robot to program it can be able to write ADA applications. On the other hand, the definition of the professional requirements and education of robot programmers is far to be reached. ADA has many chances to be a reasonable solution for programming robotics cells. The main shortcoming of this philosophy of algorithmic and explicit programming is that it is unsuitable to deal with a complicated cell in which many events may happen, time and sequence constraints can be meet by different solutions. In this case expert systems, as for instance GARI (9) seem a more flexible and understandable way of planning and controlling the cell.

FUNCTIONAL LANGUAGES AND LOGIC PROGRAMMING

After 1980 a lot of literature has stressed the idea that the future of computer languages can be different from the present in some radical way. The evolution of languages as we have so far seen from Algol to Pascal to Ada can be a dead end for computing. Those languages are based strictly on the Von Neumann architecture of computers, and that architecture is unlike to continue in the future computer generations because it has an unnecessary bottleneck in accessing memory. Languages using assignments access memory one word at a time, and assignments make user languages more suitable to the way computers operate than to the way humans think. The next generation of computers should avoid this bottleneck. Many architectural solutions are available for that, all of them relay on using functional languages. What makes a functional language attractive is its problem-oriented expression, because it expresses functions and doesn't care about memory locations, is its way to be implemented as a very restricted kernel (the function definition and composition operators) and then to grow in every way using user defined functions, it is suitability to run on largely distributed architectures because it does not produce side effects (no global memory is used). We have found an accent on functionality in many robot languages. Mini (28), LAMA-S (12), AML (29) , and Lenny (30) have provided some way to obtain functional capabilities, mainly extensibility.

Another language with full functional capabilities is *Prolog*. It is also the most successful language for logic programming. The Japanese Fifth Generation programs mostly relies on it as the basic language for future computers. Its use for robot programming has not yet been tried. In some application fields near to robots, i.e. CAD, its use has been demonstrated useful. In a comparison (18) between a 3D graphic program written in Pascal and the same program written in Prolog the Prolog implementation was more concise, readable and clear than the Pascal version. It also took less storage and ran faster than the Pascal compiled version. Since Prolog has been used for operating systems as well as for plan generation and has various ways for managing arrays it applicability to robotics waits only to be fully demonstrated. We are guessing that a prolog implementation will be accepted both by Artificial Intelligence oriented users and by mathematical oriented people.

EUROPEAN ROBOT LANGUAGES

The European scene of robot programming is very active. European was the first commercially available language for robot, SIGLA, and European are some of the most advanced projects. In the following we make a short presentation of all of them we have so far found in the literature.

HELP (8,11). It is the language developed by DEA (Italy) for their Pragma A 3000 (Allegro in USA) assembly robot. It allows concurrent programming and structured programming. The syntax is Pascal-like, all the manipulation functions are provided as subroutines. Signal and wait provide synchronization between different tasks. Any kind of sensors can be connected using a rich set of I/O ports operations. The robot is modular, different arms and different degrees of freedom for each arm can be organized. The coordinate system is cartesian, and two rotary axes can be added to every wrist. The application programs are usually provided with the installation of the robot. Major applications are in the automotive industry, electronic assembly, precision mechanics. It is implemented on DEC LSI 11 computers under the RT-11 operating system.

LAMA-S (12). The language developed by the Spartacus project, a project aimed at developing robots to help handicapped people in many every day life tasks, as serving drinks or food. LAMA-S used APL as implementation language. The user level functions are translated into a low level language, PRIMA, and then executed. Besides move instructions based on the use of frames LAMA-S provides real time primitives and parallel execution of tasks. The language uses two structures to define the execution order: sequence block, to indicate that all the instructions inside are to be executed sequentially, and parallel block, to indicate that all the instructions inside are initiated in parallel (something as cobegin-coend structure). Other standard control structures are provided. The use of APL demonstrated, according to the authors, that APL is a good implementation tool because it allows functional extensibility of the language. On the other hand they do not recommend it for industrial use because of the following shortcomings: it needs an APL machine to run, the APL syntax is not convenient, the syntax analysis is not perfect, it is difficult to implement interactive programming using APL.

LENNY (30). The language under development at the University of Genova (Italy) to be used to describe movements for an emulated anthropomorphic arm, with seven degrees of freedom. It is intended as a language powerful enough to express complex chains of actions and understandable by humans as a way to represent processes and concurrent computations. One of the key issue of Lenny is functionality. No reference can be done to any absolute kinematic quantity. References are always to actual mechanical context. In Lenny the robot reference frame is fixed in the shoulder and commands like up, down, right, etc. refer to that coordinate system. Functionality will enable Lenny to use any new procedure as part of the language. Lenny got its name from an Asimov novel in which accidentally a robot, named Lenny, became able to learn.

LM: Langage Manipulation (21,23). A language developed at the University of Grenoble (France). It is implemented on a Robitron robot (4 degrees of freedom) cooperating with a Barras robot (2 degrees of freedom), a TH8 of Renault, and a Kremlin robot, both with 6 degrees of freedom, and commercially available on the Scemi robot. It is Pascal-like and frame oriented and provides many of the features of AL but coordination and parallel execution of tasks. It is integrated with LM-Geo (24), a system used to infer bodies positions from geometrical relations. LM-Geo produces program declarations and instructions in LM. LM-Geo resembles RAPT but it doesn't use symbolic algebraic calculus to find the frames which satisfies the equations. It analitycally computes the values.

LMAC (19). A system for flexible manufacturing developed at the University of Besancon

(France). It was designed to assure a safe control of different mechanical devices in the automated cell; for that it performs many checks before actually executing code. It offers modularity based on the implementation of abstract data types, it provides generic modules (the types of data belonging to that kind of module can be specified at run time), and object parametrization. External procedures written in any language can be called by LMAC programs. Different tasks representing different real-time processes can be defined and executed. Synchronization is based on Dijkstra guarded commands. Even though its external form resembles Concurrent Pascal it has been completely rewritten in Pascal.

LPR: Langage de Programmation pour Robots (2). A language developed by Renault and the University of Montpelier (France). It is based on defining state graphs and transition conditions. Transition conditions are used also to synchronize actions. All the graphs at the same level are executed in parallel by the supervisor; every 20 ms an action from each of the same level graphs is executed. Up to 24 input/output ports can be used by LPR to provide sensor interface and synchronization with other devices. LPR runs on a VAX 11/780 and produces code for an Intel 8086 microcomputer controlling the robot. It is available on robots produced by Renault and by ACMA Robotique.

MAL: Multipurpose Assembly Language (14,15). The language developed at the Milan Polytechnic to program a two arm cartesian robot evolved from Olivetti SIGMA. It is a Basic-like system which features synchronization and parallel execution of tasks as well as movement instructions and sensor interfaces. Subroutine calls with argument lists are supported. MAL is composed by two parts, a translator from the input language into intermediate code and an interpreter of the intermediate code. The intermediate code is interfaced with a multimicro hierarchical structure, and all the joints are individually driven by different microcomputers. Force sensing is also controlled by a devoted microcomputer. Photo diodes on the fingers are used as binary sensors. Due to the mechanical architecture of Supersigma collisions between the arms are hardware detected.

PASRO: PAScal for RObots (4). It is provided by the German company Biomatik. It is based on the Pascal language which has been added data types and procedures used to perform robot specific tasks. They are stored in a library and callable by any standard Pascal compiler. It is based on the AL experience. The company may provide assistance in order to modify the coordinate transformations and the control interface for a new kind of robot. Procedures are provided to drive the arm point to point, or along a continuous path. The first implementation of PASRO has been tested on a Microrobot.

Portable AL (13). It is an implementation of the AL programming environment done at Karlsruhe University on mini and micro computers. It incorporates AL compiler (13), POINTY (17) and a debugging system. A dedicated operating system was developed to support I/O and multi-tasking. It runs on a PDP 11/34 and an LSI - 11/2 which control the PUMA 500 robot.

RAPT: Robot APT (1,25). In its actual implementation RAPT is an APT like language used to describe assemblies in terms of geometric relations and to transform them into VAL programs. A RAPT program consists of a description of the parts involved, the robot, and the workstation and an assembly plan. The assembly plan is a list of geometric relations expressing what geometrical relations should held after a step in the assembly has been done. The program is completely independent of the type of the robot used. Sensors are not integrated movements are not checked against collision avoidance. All the bodies are described as having a position (which is a frame) and some features. Features are plane, cylindrical or spherical faces. A reference system is automatically set in every feature. Against and fits are most used relations. Other relations are used to indicate translation or rotation degrees of freedom left. RAPT builds a graph of those relations and tries to reduce it to the minimum graph using a set of rules. From the reduced graph a VAL program is produced. The Computervision CADD3 system has been used to build the models and to give graphics routines.

ROBEX: ROBoter EXapt (32). The off-line programming system developed at Aachen (Germany) as a programming tool for FMS. Its main purposes are to develop APT for FMS and for robot off-line programming, and to be independent of the kind of robot used. Applications are in workpieces handling. APT style of programming is used to describe geometry, while the ROBEX extensions are robot movement instructions, interactions with sensor (now only binary ones), and synchronization with peripherals (machine tools, conveyor belts...). In this APT like system for FMS three languages will be used: EXAPT, for NC part programming, ROBEX, for part handling programs, and NCMES, for measuring programs. The system is portable in two ways: it is implemented in FORTRAN IV and it generates robot independent pseudo-code which is sent to the appropriate robot for further processing and execution. The user interactively or using a graphic interface inputs coordinates and geometry of the world and programs.

SIGLA: SIGma LAnguage (3,26). The language for programming Olivetti SIGMA robots. Now quite obsolete and under replacement, it has been available since 1975. SIGLA is a complete software system which includes: a supervisor, which interprets a job control language, a teaching module which allows teaching-by-guiding features, an execution module, editing and saving of program and data. SIGLA has been in use for years at the Olivetti plant in Crema (Italy). Its applications span from assembly to riveting, drilling, milling. All the system and the application program run in 4K of memory, and this compactness was necessary at the time SIGMA was delivered because memory was still expensive.

SRL: Structured Robot Language (6). The language under development at the University of

Karlsruhe. It is a successor of Portable AL and owns some to Pascal too. Data types as in Pascal are added to AI data types. The declaration part contains also a specification of the system components. Instructions can be executed sequentially, in parallel, or in a cyclic or delayed way. Different motions are available, in particular straight and circular motions. The project of SRL is part of a standardization project. The source SRL code is translated into an intermediate code, IRDATA, which is a machine independent code.

VML: Virtual Machine Language (16). The language developed in cooperation by the Milan Polytechnic and the CNR Ladseb of Padova (Italy). Intended as an intermediate language between Artificial Intelligence systems and robot it receives points in the cartesian space and transforms them into joint space. It manages task definition and synchronization as well. It is part of a hierarchical architecture, in which 3 levels are today implemented.

CONCLUSIONS

It is hard to not get lost in the many different languages for robots available around. We have tried to review them trying to see what experience they have provided and what open problems they have not solved. Many issues has not been addressed and we didn't expect to be complete. Most of the new commercially available languages for North American market have not been included. We wanted to make our overview on the basis of the experiences available in Europe now. While most of the attention is now focused on acquiring manipulator level systems we have tried to discover what other trends and experiences are available to expand robot programming toward more ambitious tasks.

ACKNOWLEDGEMENTS

Thanks are given to J. Bach, A. Haurat, and J. Lefebre who provided unpublished material.

REFERENCES

1. Ambler, A. P. and Popplestone, R. J., *Inferring the positions of bodies from specified spatial relationships*, Artificial Intelligence 6, pp 157-174 (1975).

2. Bach, J., *LPR Description*, unpublished, Renault, France (1983).

3. Banzano, T. and Buronzo, A., *SIGLA - Olivetti robot programming language*, Proc. Programming Methods and Languages for industrial robots, IRIA, France, pp 117-124 (1979).

4. Biomatik, *PASRO - Pascal for robots*, Biomatik Co., Freiburg, West Germany (1983).

5. Blume, C., *A structured way of implementing the high level programming language*

AL on a Mini and Microcomputer configuration, Proc. 11th ISIR, Tokyo, Japan, pp 663-674 (1981).

6. Blume, C. and Jacob, W., *Design of a Structured Robot Language (SRL)*, Proc. Advanced Software in Robotics, Liege, Belgium (1983).

7. Bonner, S. and Shin, K. G., *A comparative Study of Robot Languages*, IEEE Computer, N. 12, pp 82-96 (1982).

8. Camera, A. and Migliardi, G. F., *Integrating parts inspection and functional control during automatic assembly*, Assembly Automation, 1, 2, pp 78-82 (1981).

9. Descotte, T. and Latombe, J. C., *GARI: a problem-solver that plans how to machine mechanical parts*, Proc. 7th IJCAI, Vancouver, Canada, pp 766-772 (1981).

10. DoD, *Reference Manual for the ADA Programming Language*. Proposed Standard Document, Dept. of Defense, USA (1980).

11. Donato, G. and Camera, A., *A high level programming language for a new multiarm assembly robot*, Proc. 1st Int. Conf. on Automated Assembly, pp 67-76 (1980).

12. Falek, D. and Parent, M., *An evolutive language for an intelligent robot*, The industrial robot, 7, 3, pp 168-171 (1980).

13. Finkel, R. et al, *An overview of AL, a programming system for automation*, Proc. 4th IJCAI, Tbilisi, USSR (1975).

14. Gini, G. et al, *A multi-task system for robot programming*, ACM Sigplan Notices, Vol. 14, N 9 (1979).

15. Gini, G. et al, *MAL: a multi-task system for mechanical assembly*, Proc. Programming Methods and Languages for Industrial Robots, IRIA, France (1979).

16. Gini, G. et al, *Distributed robot programming*, Proc. 10th ISIR, Milan, Italy (1980).

17. Gini, G. and Gini, M., *Interactive development of object handling programs*, Computer Languages, Vol. 7, N. 1 (1982).

18. Gonzalez, J. C., Williams, M. H., Aitchison, D. E., *Evaluation of the effectiveness of Prolog for a CAD application*, IEEE CG&A, N 3, pp 67-75, (1984).

19. Haurat, A. and Thomas, M. C., *LMAC: a language generator system for the command of industrial robots*, Proc. 13th ISIR, Chicago, Illinois, pp 12-69 (1983).

20. Kempf, K., *Artificial Intelligence Applications in Robotics - A Tutorial*, IJCAI 83 Tutorial, Karlsruhe, Germany (1983).

21. Latombe, J. C. and Mazer, E., *LM: a high-level programming language for controlling assembly robots*, Proc. 11th ISIR, Tokyo, Japan, pp 683-690 (1981).

22. Lozano-Perez, T., *Robot programming*, Proc. of the IEEE, 17, 7 (1983).

23. Mazer, E., *Geometric programming of assembly robots*, Proc. Advanced Software in Robotics, Liege, Belgium (1983).

24. Miribel, J. F. and Mazer, E., *Manuel d'utilisation du langage LM*, Research report IMAG, University of Grenoble, France (1982).

25. Popplestone, R. J. et al, *An interpreter for a language for describing assemblies*, Artificial Intelligence, Vol 14, pp 79-107 (1980).

26. Salmon, M., *SIGLA - The Olivetti SIGMA robot Programming Language*, Proc. 8th ISIR, Stuttgart, Germany, pp 358-363 (1978).

27. Schimano, B., *VAL: an industrial robot programming and control system*, Proc. Programming Languages and methods for industrial robots, IRIA, France (1979).

28. Silver, D., *The littler robot system*, MIT Artificial Intelligence Lab. Rep. AIM 273 (1973)

29. Taylor, R. H., Summers, P. D., and Meyer, J. M., *AML: A Manufacturing Language*, The International Journal of Robotics Research, 1, 3, pp 19-41 (1982).

30. Verardo, A. and Zaccaria, R., *Lenny Reference Manual (in Italian)*. Internal Report, University of Genova, Genova, Italy (1982).

31. Volz, R. A., Mudge, T. N., Gal, D. A., *Using ADA as a programming language for robot-based manufacturing cells*, RSD-TR-15-83 Dep. E&C Eng., The University of Michigan, Ann Arbor, Michigan (1983).

32. Weck, M. and Eversheim, E., *ROBEX - An off line programming system for industrial robots*, Proc. 11th ISIR, Tokyo, Japan, pp 655-662 (1981).

33. *CAM-I proposes standards in robot software*, The Industrial Robot, pp 252-253 (1982).

Robot Programming Using a High-Level Language and CAD Facilities

C. LAUGIER*

LIFIA - IMAG - BP 68 - 38402 ST-MARTIN D'HERES CEDEX

Summary

A high-level robot programming language constitutes a general purpose interface for accessing the basic functional capabilities of a robot. On the
other hand, CAD facilities give the possibility of using a subset of these
capabilities in an easier fashion. In this paper, we show how a robot programming language and CAD facilities can be combined to obtain a robot programming system satisfying the need for generality, and allowing an easy
connection with the basic robot programming functions. Such a connection is
based on a "complete" simulator providing facilities for executing robot
control programs on a graphic display, for "emulating" sensors during this
execution, and for interactively guiding the robot on the display during
programming by showing phases. Our proposal is illustrated by a system developped at the LIFIA/IMAG laboratory, using the LM robot programming language
designed and implemented by our group .

1. INTRODUCTION

A high-level robot programming language constitutes a general purpose interface for accessing the basic functional capabilities of a robot. In particular, it allows to describe any manipulation task in terms of motions and
actions to be performed by the robot end-effector. Such a programming pattern
features a high degree of diversification at a comparatively low procurement
cost (a micro-computer may be sufficient for supporting the operating system). Nevertheless, the relevant robot control programs can hardly be
written and debugged as they handle data requiring a good control of the
physical world. This is the reason why it is necessary to have suitable
tools designed for facilitating the programmer's job by giving a more understandable picture of the data handled by the program.

CAD facilities give the possibility of using some basic functional capabilities of the robot in a simpler fashion. In particular, they provide the
user with powerful graphic tools allowing off-line programming. Such facilities may be easily implemented on a medium capacity computer connected
with a CAD data base and an interactive graphic display terminal. But,

*Researcher at INRIA

various graphic simulation levels may be considered for implementation according to the data processing capabilities available for use by the operator [1]. The graphic outputs produced by the related systems may vary from a small set of pictures for representing the task, to a dynamic simulation coupled with a high degree of interactivity. However, in spite of all graphic improvements, the simulation remains unrealistic since the progress of a manipulation program varies from an execution to another one. These variations are jointly caused by the imperfections existing within the mechanical structure, the uncertainties connected with the manipulated objects and the mechanical faults such as slipping effects. Within the program, such variations are detected and processed using sensor data. Consequently, a realistic simulation of robot control programs cannot be obtained without an elementary sensor management.

In this paper, we show how a robot programming language and CAD facilities can be combined to obtain a robot programming system satisfying the need for generality, and allowing a simple connection with the basic robot programming functions. Such a connection is based on a "complete" graphic simulator providing the user with facilities : (1) for executing robot control programs on a graphical display, (2) for "emulating" the related simulation processes according to external data (for example : data provided by sensors or random perturbations introduced into the world model), and (3) for interactively guiding the robot on the graphic   display during programming by showing phases. Our proposal is illustrated by a system developped at the LIFIA-IMAG laboratory. This system makes use of the LM language for programming and controlling real robots (LM is a robot programming language designed and implemented by our group [2,3]). It is distributed on two different computers : the graphic simulator and the CAD data base are currently located on a CII-HB-68 computer, and the LM interpretor is running on a LSI-11/23 microcomputer controlling a six degrees of freedom SCEMI robot. Graphical results are either displayed as color drawings or as shaded pictures. In the last case, the system makes use of the HELIOS specialized graphical terminal [6] designed by the graphic group of the IMAG Institute.

2. MANIPULATION-LEVEL SYMBOLIC PROGRAMMING

A high-level robot programming language allows to describe any manipulation task in terms of motions and actions to be performed by the robot end-effector. For this purpose, it includes basic data structures and instructions set one uses to find in conventional language, along with new constructs suitable for specifying robot tasks. The most important of these constructs deal with world modelling, describing end-effector motions and operations, sensor data gathering, and synchronizing the robot with external processes like part feeders or other robots. All manipulation languages make it possible to describe tasks within a cartesian world made up of reference frames. This programming method basically consists in modelling the robot end-effector and the objects within the surrounding space by means of a set of reference frames properly selected. The programmer then acts in terms of the respective positions and orientations of such frames, independently of the shape of the manipulated objects and of the mechanical configuration of the manipulator. He has available for this purpose suitable types of data such as frames, vectors and transforms. He also uses statements adapted to the processing of these data, for instance the cross-product of vectors, or the extraction of a transform. Within a manipulation program, the geometry of the task is sketched by cartesian frames attached to the parts to be assembled. Then, each individual object of the world is modelled through a set of linked frames including a <u>locating frame</u> stated in a fixed reference frame named BASE or STATION, and various <u>functional frames</u> stipulated for the purpose of specific manipulation operations (for instance grasping operations). The locating frame makes it possible to locate at any time the object within the robot workspace. It may be modified automatically by the language interpreter during a motion, or by the program using a statement of the type :

loc-object := reference*transform

where "reference" designates the reference frame of the working space, and "transform" refers to the geometrical transformation that makes possible to convert from the reference frame to the object locating frame. Likewise, any functional frame may be added to the model by using a statement of the type :

ATTACH new-frame TO frame-i USING transform

where "frame-i" designates an existing frame.

This symbolism makes it easy to describe any robot motion by an instruction of the type :

MOVE frame1 TO frame2 UNTIL condition

where "frame1" is a moving frame (namely a frame associated with the end-effector or a frame momentarily affixed to this latter by a specific program statement), "frame2" denotes the goal position and "condition" represents a boolean condition likely to stop the motion. At the end of this

motion, "frame1" should lie in the position specified by "frame2", unless
the boolean condition which includes sensor data has become true during the
performance of the motion.

Today, a lot of manipulation languages exist [20] and some of them are com-
merciably available : LM (ITMI)   , VAL (Unimation)   , AML (IBM)   ,
RAIL (Automatix)   ... These languages have been used on a wide variety
of manipulation tasks including arc welding and mechanical assembly [21].
However, programming a robot using such a method remains a difficult task
requiring a good control of the physical world : the programmer must state
quantitatively all robot motions and make sure that no collision is likely
to occur ; he also has to manage the data provided by sensors with a view of
matching the robot behaviour to the world uncertainties. This is the reason
why various programming aids have also been designed. For example, POINTY
[22] provides interactive functions for constructing programs expressed in
the AL language using guiding techniques. More recent languages like RAPT
[23] and LM-GEO [24] make use of simple geometric models of objects for defining frame positions in terms of symbolic geometric relations rather than
by quantitative transforms (see [25]). Graphic simulators also have been developped as off-line programming tools. The state of the art in this domain is
presented in the next section.

3. STATE OF THE ART IN GRAPHIC SIMULATION OF ROBOTS

Several graphic simulation systems have been developped, or are presently
under development. Some of them work in a two-dimensional world ; they
allow to simulate robots having three or four degrees of freedom using a
personal computer and a low cost graphical terminal, but these systems make
use of very simple graphic representations to visualize the robot and its
workspace. This is the case, for example, of the SIM/7535 system developped
at the IBM research center [7] and of the system designed by the Tokyo
University to simulate a four axis Sankyo robot programmed using a specific
robot language called EARLS-2 [8].

Other simulators operate in the three dimensional world. Most of them are
based on existing CAD systems : CATIA [9,10], GEOMAP [11], SAMMIE [12,13],
GDP [14]. Nevertheless, the graphic capabilities vary widely from one implementation to another. For example, the simulator designed by the Tokyo
University requires several seconds of CPU for computing a wire-frame picture [11], while the robot function of the CATIA system [10] is able to operate on an almost realtime basis with simple pictures displayed with hidden

lines removal. In order to produce faster graphics, several systems use poor graphic representations based on parallelepipeds and cylinders. The simulator developped by the WZL laboratory [15] is an example of this kind of system. Another approach consists in solving the real-time problem by an adequate hardware configuration. This approach has been used at the Mitsubishi research laboratory where a system made of several modules has been implemented on a high speed hardware built with 8086 and 8087 micro-computer cards [16].

Very few simulators have been developped using an existing robot control language. The main research laboratories working in this way are the IBM research center, the Stanford University, and the LIFIA-IMAG laboratory of Grenoble. IBM's EMULA system [14] is based on the AML language:; it allows to simulate the IBM-7565 robot on the IBM-370 computer. The simulator developped at Stanford University makes use of a subset of the AL language to simulate a PUMA-500 robot modeled with the ACRONYM system [17,18]. Finally, the simulator implemented in our laboratory was designed as an extension of the LM system. It is important to notice that, among all these simulators, only the IBM's EMULA system and the LIFIA-IMAG system offer the capabilities required to achieve a sensor managing job.

New work concerning Computer-Aided-Design of robots and graphic programming is currently in progress in several laboratories. Significant results have already been obtained at the LAM laboratory [10], but no real connection with an existing robot control language has still been completed.

4. OFF-LINE PROGRAMMING CAPABILITIES

The off-line programming capabilities available in the system described in this paper are provided by a graphic simulator allowing to simulate any robot control program, and to interactively enter various trajectories. This simulator needs three types of data (see figure 1) :
- the geometric model of the objects and the kinematic model of the robot,
- the manipulating task to be simulated,
- physical events such as perturbations which cannot be described by the world model.

The data related to the geometric model are potentially present in CAD data bases. They include a polyedral representation associated with every solid object lying in the three dimensional scene. Kinematic type data are given by specialized functions designed for mechanism modelling. In the current implementation of our system, these data are obtained using particular functions of the SMGR geometric modeller [5].

Data concerning the task to be completed are derived from a manipulation program, and are provided by the interpreter of the robot programming language. Data concerning the physical world are introduced into the system by the operator. This approach allows the user to _emulate_ the simulation process by modifying some external parameters. These parameters are the time basis, data provided by sensors and physical events represented by random perturbations introduced intċ the world model.

All previous data are used by the system for building and updating a structured model of the robot world. It is this model which varies during the simulation process, and which is displayed by the graphic functions of the system. Once equiped with this data base, the simulator has the capabilities for executing the manipulation program statements on the graphic terminal and for recording numerical informations concerning specific key positions and trajectories. In the first operating mode, the simulator controls the robot's kinematics according to the data furnished by the language interpreter. These data are denoted by vectors in terms of joint coordinates. In the other mode, the robot is manually controlled by the operator by means of an interactive device such as a joystick or a function keyboard. He then works in the learning mode, for the purpose of being able to enter the key positions which are required for carrying out the programmed task. To accomplish this, the operator may either activate each robot joint individually, or directly move the end-effector in the cartesian space using some specialized functions of the type of those introduced into the CATIA system [9].

5. INTERACTIVE COMMANDS AVAILABLE IN THE SYSTEM

All the commands addressed to the system may be directly inserted into the manipulation program, or given interactively by the operator. In the first case, the user obtains after each simulation a dynamic representation of the whole manipulation task. In this way, he can study the behavior of his program with respect to different physical situations by executing it in several world configurations. In the other case, the operator works "on line" with the simulation process in order to debug or to complete his manipulation program. Thus, he may stop the execution of the task at any time in order to modify the simulation context (world model, graphic parameters, external parameters), or to enter new subtasks using the specialized graphical

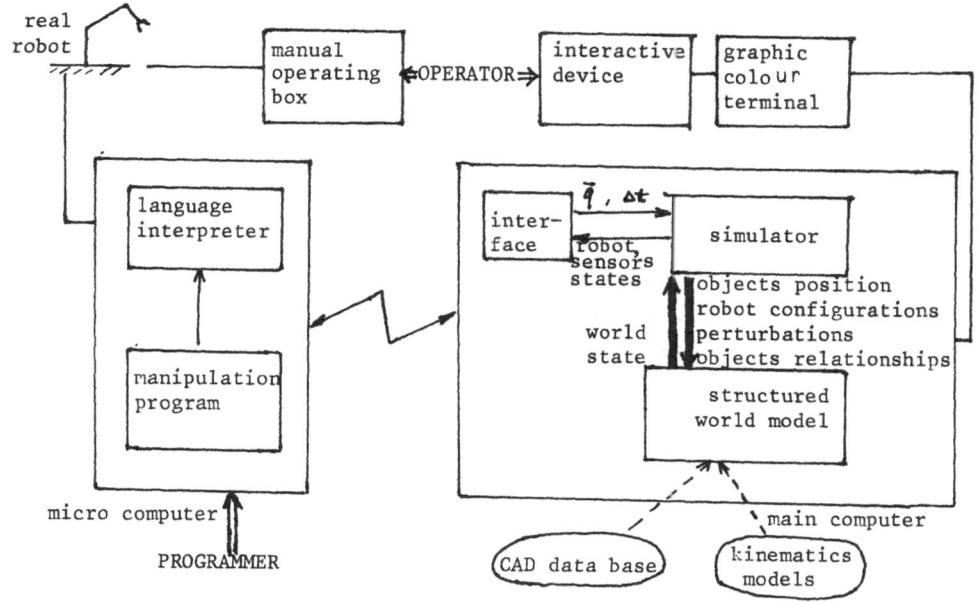

<u>Figure 1</u> : Functional structure of the system

functions of the system (see section 8). He may also make use of geometrical functions for interference checking, when pictures provides by the system do not allow a visual control of collisions.

The commands available in the system are grouped into three sets allowing respectively to control the simulation operations, to activate the emulation functions, and to program trajectories on the graphic display. These commands are roughly described in the next sections. A more detailed description of the whole facili ties provided by the LIFIA-IMAG simulator may be found in [19].

6. SIMULATION FACILITIES

The simulation commands allow the user to interact with the world model and to control the displaying operations.

Three types of operations may be used for modifying the world model : (1) placing a new object into the robot workspace, (2) moving any componant of the world (objects or robot links), and (3) modifying the world relational structure. Commands associated with the operations (1) and (2) are used to specify the dynamics of the three dimensional scene. Other commands may be used to define temporary or permanent physical constraints applied to particular objects of the world. They lead to the manipulation of "rigid structures" representing physical affixments created by some robot actions. Such situations arise when mechanical parts have been assembled, or when an object or a tool has been grasped by the robot end-effector. The displaying operations are executed in connection with a visual model including : (1) a description of the scene lighting, (2) a representation of the visual aspect of objects, and (3) a caracterization of the observer parameters. This model is recorded in the graphic data base. It may be modified during the simulation process using some specific commands of the system. In the LIFIA-IMAG simulator, data concerning the scene lighting and the visual aspect of objects are mainly processed by the HELIOS terminal [6]. These data are not considered when the displaying operations are executed on a classical graphic terminal.

The observer parameters are used by the simulator to determine the components of the three dimensional scenes to be displayed, and for controlling the required camara movements. In order to have a natural mode of representation of these parameters, the simulator maintains a model of the viewing camera as being on a sphere centered on a point of interest (called the "vision center"). The latitude and the longitude of the camera may easily be changed by the user, as well as the focal length (see [4]). This approach makes graphic modifications easy to describe. It also allows to fix the vision center onto a mobile object in order to automatically follow it during its motion. For example, the user may apply a zoom centered on the robot gripper in order to maintain a detailed view of the gripper in the center of the screen during all the movement.

7. EMULATION FUNCTIONS

The simulation process may be emulated in three different ways : by modifying the time basis, by introducing random perturbations into the world model, and by simulating the sensors behavior.

The time parameters are used to define the basic internal cycle of the simulator, and to specify the frequency of the displaying operations. The world model is then updated at each cycle, and it is displayed at intervals defined as a multiple of this basic cycle. Using this method it is possible to obtain sets of pictures for an animated film by setting the display rate to 24 frames/second, and by choosing a basic cycle equal to the one used by the interpreter of the robot programming language.

The perturbations are applied to the designed objects or to the robot links using a random function. The range of values provided by this function may be modified interactively by the operator.

The sensor simulation may be realized in two ways : either by manually entering the values which are to be set to the concerned sensors, or by using specialized geometrical functions. In the first operating mode, the user is in charge of stopping the guarded commands for simulating that a threshold value is reached, or that a boolean expression including sensor data is true. In the other mode, the sensor management is made using geometrical functions based on the primitive interference checking provided by the system . As in the EMULA system [14], these functions may be associated dynamically with sensors. Then, at each cycle, all associated functions are invoked, and the results are assigned to the corresponding sensors.

Two types of functions may be used for sensor simulation : the contact function and the local analysis function. The first one can be used to detect any contact between several objects given as arguments to the function. The local analysis function can be used to determine the presence or the absence of an object in a particular area of the space. This area, which is either cylindrical or conical, represents the light beam produced by a proximity sensor.

8. GRAPHIC PROGRAMMING FACILITIES

Programming facilities provided by the specialized graphic functions of the system are similar to those available in a geometric relation-based programming language [25] : they are aimed at acting as an interface between a manipulation language and a geometric data base. Such programming facilities make reasoning in the three dimensional world easier, through providing the user with the capability of denoting the positions taken by the robot in terms of elementary geometric relations. Robot control quantification is then performed by the system which generates conventional manipulation statements expressed in an existing manipulation language.

In graphic programming the operator stipulates graphically a target position for each individual robot motion. To accomplish this he uses elementary geometric relations of the following type :

"MOVE entity A of robot ONTO entity B of object O"

Each relation described is immediatly interpreted by the system. This makes the robot entity to be moved towards the computed target position. When no additional constraint is entered by the operator, this target position is

determined using the minimum displacement rule which leads to minimize the
rotational and the translational component of the computed transformation.
In this case, the trajectory followed by the robot end-effector when executing the commanded motion is computed by the language interpretor in connection with the moving mode stipulated by the operator (free mode, cartesian
mode...). Another possibility would consist in describing graphically the
whole trajectory in terms of curves to be tracked. Such a facility is currently available in the LM language, and it will be soon added to our
off-line programming system.

The geometric model required for graphic programming includes a conventional
graphic representation of objects (for the displaying operations), and a
functional component allowing the description of the geometric relations to be
achieved.This component is made up of elementary geometric entities attached
either to the robot end-effector, or to the objects belonging to the world
model. These entities are either points, straight lines or planes. They
may be created, deleted or momentarily invalidated by the user during the
programming process. For instance, the operator may define a new straight
line in the model by designating two existing points. He may also remove
from the screen all irrelevant graphic information in order to obtain a
more understandable picture of the data handled by the system during the
current programming phase.

The off-line programming facilities provided by our system are potentially
similar to those implemented in the CATIA system [10]. But, resulting informations concerning the real robot motions and actions are not on the same
command level : in the CATIA system the interpretation of a subtask leads
to generate simple trajectories for the robot end-effector ; in our case,it results
in the addition of new statements to the manipulation program under construction (see figure 2). This symbolic approach is more general and more flexible because it allows combination of the expressional power of a symbolic language with the versatility of CAD facilities. Moreover, it intrinsically brings
elements of solutions   to the major problem of off-line programming :
the discrepency existing between the simulation models and the real processes.

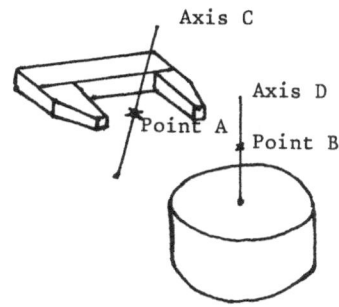

Axis C

Axis D

Point A   Point B

geometric relations :   { Point A <u>onto</u> Point B
                             Axis C <u>onto</u> Axis D

Interpretation : { CATIA + coordinate transformer of the real robot →{c1,c2..cn}
                    LIFIA simulator + LM interpreter   →{T:=(....);
                                                          MOVE ROBOT TO ROBOT*T;

Figure 2 : Interpretation of geometric relations

9. WORLD MODELLING

In order to provide the user with the CAD facilities described above, the
system must manage a world model including : (1) a polyhedral representation
for solid objects, (2) a caracterization of the spatial relationships and
constraints existing between the world items, (3) a specification of the
functional geometric entities created for graphic programming, and (4) a
kinematic representation for the various mechanical mechanisms.

9.1. The world model

Each solid object belonging to the world model and each robot component is
represented by a cartesian frame associated with a set of polygonal faces.
The cartesian frame allows the system to localize every object in the robot
workspace, and the sets of faces are used for the displaying operation. The
structure of the world is caracterized by a graph in which each arc denotes
a particular relationship between two world items, and each node represents
an object of the world. Data supported by arcs are constant or parametrized
geometrical transformations. Data associated with nodes are either single
cartesian frames (for fictitious objects or objects made of several sub-
parts), or lists of faces (for solid objects).

Three kinds of structure are used in the world model [19] : (1) a hierarchical type structure inferred from spatial relationships existing between
objects, (2) a tree structure associated with each kinematic mecanism, and
(3) a graph structure characterizing permanent or temporary "rigid structures".
This multiple structure allows the system to apply several methods suitably
designed to update the world model. For example it is more convenient to follow the tree structure associated with a mecanism, when kinematic computations are to be completed (see figure 3).

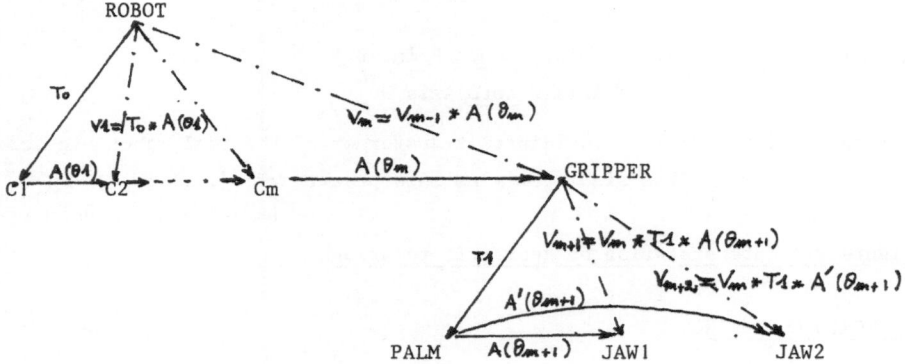

Figure 3 : Evaluation of the tree structure associated with the robot
mecanism.

9.2. The kinematic model of the robot

The robot is represented by a "kinematic open loop" made of n solid components linked together by joints. Two basic types of joints are used for this
description : the <u>revolute</u> joint which allows a relative rotational motion
around one axis, and the <u>prismatic</u> joint which allows a relative translational motion along an axis.

More complex joints may also be described using these two basical types of
joints. For example, a spherical type joint having three degrees of rotational freedom may be represented by combining three revolute joints with
converging axis. In this case, two fictitious parts represented by cartesian
frames are introduced by the system into the kinematic chain. Other mecanisms
involving several parts lonked together by a single joint may be described
using two special types of joints : the <u>poly-revolute</u> joint and the <u>poly-
prismatic</u> joint. These new types of joints, which are single expansions of
the basic one described above, allow to distribute a single moving command
between several parts. For example, this situation arises when moving the
jaws of the gripper.

In order to represent robots having elementary kinematic loops (parallelogram), the system makes use of <u>c-revolute</u> type joints for coupling the "pas-
sive joints" of the loop with the "active" revolute joint controlling the
mechanism (see figure 4). This approach, which is similar to the one employed
in the CATIA system [9] permits to keep a tree structure for those mechanisms
without artificially increasing their number of degrees of freedom.

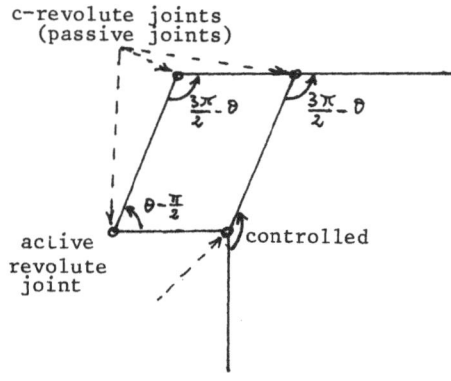

Figure 4 : Parallelogram type structure

9. CONCLUSION

The system described in this paper offers the user of industrial robots a
powerful tool for developping and debugging robot control programs : it
accepts the same language as the real robot along with CAD facilities for
specifying simulation and emulation operations and for interactively guiding the robot on the graphic   display. Moreover, its sophisticated graphic
capabilities coupled with a high degree of interactivity make the control
of collisions easy. An important point with regard to industrial applications is the fact that the system is distributed on   two different computers in order to provide the user with both "on-line" and "off-line" programming facilities. In the current implementation, the graphic simulator and
the CAD data base are located on a CII-HB68 computer and the LM runtime system is running on a LSI11/23 microcomputer controlling a six degrees of freedom SCEMI robot. The displaying operations, which are made using general
graphic functions belonging to the LISP3D system [4] takes between 4 or 5
seconds of CPU time for computing a complex picture having about six hundreds
polygonal faces (see annexe). In the future implementation, three dimensional graphic computations will be executed on a specialized processor located between the host computer and the graphic terminal. This processor which
is currently under development, will improve the graphic capabilities of the

system. It would lead to work on an almost real-time basis, by executing the
displaying operations at a frequency of several pictures per second.

REFERENCES

[1] - C. LAUGIER
"Combining a high-level language and graphics for robot programming",
Colloque Automation et Robotique, Helsinki, Oct. 83.

[2] - J.C. LATOMBE - E. MAZER
"LM : a high-level programming language for controlling assembly robots",
11th ISIR, Oct. 81, Tokyo.

[3] - J.F. MIRIBEL - E. MAZER
"Manuel de référence LM", IMAG, Grenoble, Oct. 82.

[4] - C. LAUGIER
"LISP3D : logiciel pour la manipulation et la visualisation de scènes tri-
dimensionnelles", IMAG, Grenoble, RR n° 328, Sept. 82.

[5] - J. PERTIN-TROCCAZ
"SMGR : un système de modélisation géométrique et relationnelle pour la
robotique", IMAG, Grenoble, RR n° 422, (to appear).

[6] - F. MARTINEZ - F. FERREIRA
"HELIOS : terminal interactif pour la synthèse d'images réalistes", congrés
AFCET-TTI, Gyf sur Yvette, 1981.

[7] - J. MEYER - R. JAYARAMAN
"Simulating robotic applications on a personal computer", Computer in Mechanical Engineering, July 1983.

[8] - T. ARAI
"A robot language system with a colour graphic simulator", Proc. of Advanced software in Robotics, Liège, May 83.

[9] - A. LIEGEOIS - E. DOMBRE - P. BORREL
"Développement d'un système de CAO et de simulation de robots manipulateurs",
premières journées ARA, Poitiers, Sept. 82.

[10] - P. BORREL - F. BERNARD - A. LIEGEOIS - D. BOURCIER - E. DOMBRE
"The robotics facilities in the CAD-CAM CATIA system" edited by B. Rooks,
IFS publications, 1983.

[11] - T. SATA - F. KIMURA - A. AMANO
"Robot simulation system as a task programming tool", 11th ISIR, Tokyo,
Oct. 82.

[12] - W.B. HEGINBOTHAM - M. DOONER - K. CASE
"Rapid assessment of robot performance by interactive computer graphics",
Proc. of the 9th ISIR, Washington DC, 1979.

[13] - M. DOONED - N.K. TAYLOR - M.C. BONNEY
"Planning robot installations by CAD", Computer Aided Design Conference,
Brighton, March 82.

[14] - J. MEYER
"An emulation system for programmable sensory robots", IBM J. Res. Develop,
Vol. 25, N° 6, Nov. 1981.

[15] - D. Zuhlke
"Graphic simulation of robot actions",
WZL TH Aachen, Germany, 1983 (unpublished).

[16] - Y. TSUJIDO - N. KODAIRA -M. OSHINA
Realtime motion simulation of robots", Proc. of ICAR 83, Tokyo, 1983.

[17] - B.I. SOROKA
"Debugging robot programs with a simulator", CADCAM-8 Conference, Anaheim,
California, Nov. 80.

[18] - B.I. SOROKA
"A robot simulator", submitted to Robotic Research, January 82.

[19] - C. LAUGIER - C. EVIEUX - J. PERTIN-TROCCAZ
"Un systeme de simulation graphique de robots incluant une gestion élémen-
taire des incidents et des capteurs", IMAG, Grenoble, RR n° 421, March 84.

[20] - J.C. LATOMBE
"Survey of advanced general-purpose software for robot manipulators",
IMAG, Grenoble, RR n° 330, Novembre 82.

[21] - R.P. PAUL
"WAVE : A model-based language for manipulator control", The Industrial
Robot, March 1977.

[22] - G. GINI - M. GINI
"Object description with a manipulator", The Industrial Robot, March 78.

[23] - R.J. POPPLESTONE - A.P. AMBLER - I. BELLOS
"RAPT : a language for describing assemblies", The Industrial Robot, Sept.
78.

[24] - E. MAZER
"LM-GEO : geometric programming of assembly robots", IMAG, Grenoble, RR
n° 296, March 82.

[25] - C. LAUGIER
"La programmation des robots : expression textuelle et expression graphique"
IMAG, Grenoble, RR n° 387, July 83.

Acknowledgements

The work presented in this paper was supported in part by French ARA project
(Automatique et Robotique Avancées) and by ADI (Agence de l'Informatique).
The new developments concerning the specialized graphic processor are realized in collaboration with the ITMI company (Industrie et Technologie de
la machine Intelligente).

ANNEXES

1. Simulation of an LM program

```
PROGRAM FILM1;
TRANSFORM T1, T2, T3, T4;
FRAME PLANK1, PLANK2, CYLINDER1
      ,CYLINDER2, SUPPORT;
FRAME GRASP_P1, GRASP_P2, GRASP_C1, GRASP_C2;
FRAME ASSEMBLY_P1, ASSEMBLY_P2, ASSEMBLY_C1, ASSEMBLY_C2;
FRAME LOCATION_P1, LOCATION_P2, LOCATION_C1, LOCATION_C2;

FILE SIMUL;
FILE INPUT SUPPORT;
FILE INPUT OBJECT1;
FILE INPUT OBJECT2;
BEGIN
```

```
WRITE "PLACE PLANK1 T VX -95. T VY 330. T VZ 103." INTO SIMUL;
WRITE "PLACE PLANK2 T VX 165. T VY 460. T VZ 103.";
R VZ 90. INTO SIMUL;
WRITE "PLACE CYLINDER1 T VX 75. T VY 410. T VZ 103." INTO SIMUL;
WRITE "PLACE CYLINDER2 T VX 175. T VY 410. T VZ 103." INTO SIMUL;
WRITE "PLACE WAGON T VX -27. T VY 290.5 T VZ 130.5";
WRITE "CAMERA RH -60." INTO SIMUL;
WRITE "IMAGE" INTO SIMUL;
WRITE "PAUSE 4." INTO SIMUL;
WRITE "TRAVELLING RH 30". INTO SIMUL;
WRITE "PAUSE 2." INTO SIMUL;
FIX SPEED TO 0.2;
CO
CO      MANIPULATION OF THE FIRST PLANK
CO;
        READ SUPPORT,T1,T2 INTO SUPPORT;
        PLANK1:=SUPPORT*T2;
        LOCATION_P1:=SUPPORT*T1;
        READ T1,T2,T3 INTO OBJECT1;
        ATTACH GRASP_P1 TO PLANK1 BY T1;
        ATTACH ASSEMBLY_P1 TO PLANK1 BY T2;
        ATTACH LOCATION_P2 TO PLANK1 BY T3;
CO GRASPING;
        MOVE ROBOT TO GRASP_P1*TRANSLAT(VZ,80.);
        MOVE ROBOT TO GRASP_P1;
        OPEN GRIPPER TO 40.;
        ATTACH GRASP_P1 TO ROBOT;
        WRITE "GRASP PLANK1" INTO SIMUL;
        MOVE GRASP_P1 OF TRANSLAT(VZ,60);
CO ASSEMBLING;
        MOVE ASSEMBLY_P1 TO POSAGE_P1*TRANSLAT(VZ,60.);
        MOVE ASSEMBLY_P1 TO POSAGE_P1;
        OPEN GRIPPER TO 50.;
        DETACH ROBOT OF PRISE_P1;
        WRITE "RELEASE" IN SIMUL;
        WRITE "PAUSE1." INTO SIMUL.
        WRITE "CAMERA RH 40." INTO SIMUL;
        WRITE "CENTERV WAGON"INTO SIMUL;
        WRITE "IMAGE" INTO SIMUL;
        WRITE "PAUSE 1." INTO SIMUL;
        WRITE "TRAVELLING Z 3.0 3" INTO SIMUL;
        MOVE ROBOT OF TRANSLAT(VZ,60);
```

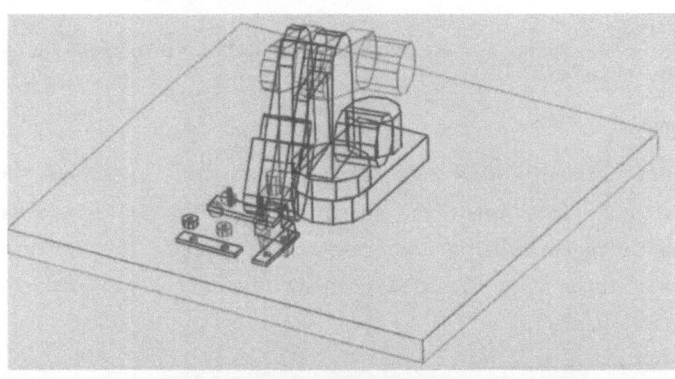

2. Graphic programming

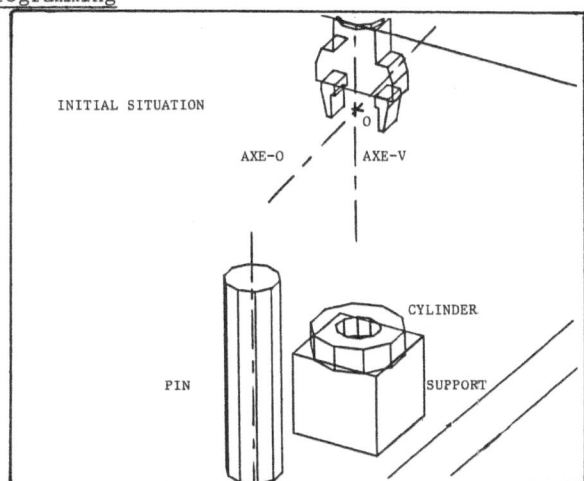

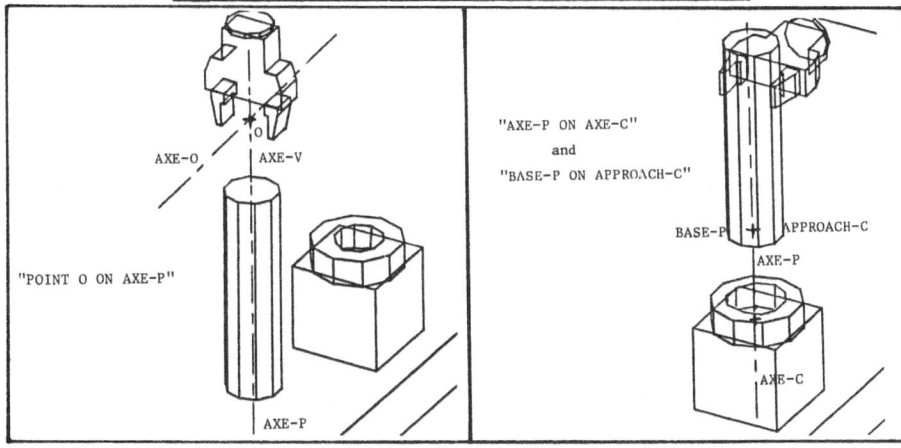

```
PROGRAM GRAPHIC-TASK;
FRAME TABLE;
FRAME SUPPORT;
FRAME CYLINDER;
FRAME PIN;
TRANSFORM T-ROBOT;
TRANSFORM T-TABLE;
TRANSFORM T-SUPPORT;
TRANSFORM T-CYLINDER;
TRANSFORM T-PIN;
BEGIN
ROBOT:=STATION*(0.0,-1.0,0.0,1.0,0.0,0.0,0.0,0.0;1.0,-400.0172,1.8261839e-03,354.99194);
TABLE:=STATION*(1.0,0.0,0.0,0.0,1.0,0.0,0.0,0.0,1.0,0.0,0.0,0.0);
SUPPORT:=STATION*(0.0,1.0,0.0,-1.0,0.0,0.0,0.0,0.0,1.0,-450.0,100.0,0.0);
CYLINDER:=STATION*(0.0,1.0,0.0,-1.0,0.0,0.0,0.0,0.0,1.0,-450.0,100.0,100.0);
PIN:=STATION*(0.0,1.0,0.0,-1.0,0.0,0.0,0.0,0.0,1.0,-400.0,250.0,0.0);
T-ROBOT:=(0.0,0.0,-1.0,0.0,1.0,0.0,1.0,0.0,0.0,-250.0,3.29303e-02,-155.0191);
MOVE ROBOT TO ROBOT*T-ROBOT;
T-ROBOT:=(1.0,0.0,0.0,0.0,1.0,0.0,0.0,0.0,1.0,9.27686e-03,-399.95667,-250.06935);
ATTACH PIN TO ROBOT BY T-ROBOT;
CLOSE GRIPPER;
T-ROBOT:=(1.0,0.0,0.0,0.0,1.0,0.0,0.0,0.0,1.0,-99.976611,-50.026069,149.98817);
MOVE ROBOT TO ROBOT*T-ROBOT;
OPEN GRIPPER;
DETACH PIN OF ROBOT;
END;
```

Implicit Robot Programming Based on a High-Level Explicit System and Using the Robot Data Base RODABAS

Christian Blume

Institute for Informatics III
Research Group: Process Control Computer Technology
Prof. Dr.-Ing. U. Rembold
University of Karlsruhe
7500 Karlsruhe 1
Federal Republic of Germany

Introduction

Especially in the field of assembly developing a plan for executing a complex task is quite a difficult job for experienced humans. Therefore programming of industrial robots for assembly needs as much support by hardware and software tools as possible. While the teach-in method has been used for years for industrial applications now textual programming is in the beginning of a widespread use. Most of these programming languages for industrial robots are of lower level and so called explicit languages. This means that all moves of the robot and all needed positions and orientations are specified explicitly by the programmer. The future goal is the development of an implicit robot programming system where the programmer specifies a complete task, e.g. assemble part A with plane bottom to part B with plane PFS, and the system generates all needed robot actions and data. The base of these new implicit systems is a powerful explicit programming kernel which provides all facilities for robot control.

This short report will give an overview about the explicit programming system with the robot language SRL and its interfaces to implicit programming and robot control level. As a first step to an implicit system the robot data base RODABAS was implemented for a world model which contains all needed data for the planning of robot actions.

1. General Requirements for Robot Programming Languages

For the automation of assembly the robot system must be equipped with sensors and interfaces to other machines or tools. Therefore the robot language should include facilities to define data structures and input/output actions with sensor interfaces, both digital/analog interfaces.

One of the essential features of a computer language for a robot is programming of the trajectory, For this purpose it is necessary that the programmer has the tool to enter the start and end positions of a movement and of the trajectory. In general, it is possible to describe textually any trajectory of a given coordinate system with the aid of a sequence of points. However, it is very difficult for a human operator to visualize this series of points in a 3-dimensional space and to describe the coordinates of each point via a programming language. In order to obtain accurate parameters, the exact location of a point must be found with the aid of a measurement. This, however, is very time consuming and awkward. In practice, the problem is solved by leading the robot's effector through its desired path and by reading the coordinates of the corresponding arm joints at predetermined points along the trajectory. The paramenters of these points are entered into the robot memory. The robot path is then reconstructed by the compiler with the aid of an interpolation algorithm. In addition to the parameters of the trajectory, the control system of the robot must have information about the orientation of the effector. This parameter can also be entered into the program by the teach-in-by-showing method.

Modern explicit programming languages for robots use the "frame" concept. With this method, a spacial point is described by a position vector with respect to a standard coordinate system. The orientation of the gripper is described by a rotation. Thus a frame consists of a position vector and an orientation, Fig. 1. When the effector is described by a position and an orientation, the following convention may be useful, Fig. 2:

- the endpoint of the position vector is located in the middle of the center line which connects the two gripper jaws,
- the gripper points into the direction of z-axis of the *frame* coordinate system,
- the y-axis of the effector runs through the gripping points of both jaws.

When one compares the assembly instructions for a human operator with programming instructions of a computer language, the following factors are of interest:

- The transfer of the dimensions from the drawing to the actual assembly object is done by the operator by looking at the drawing and by translating this information directly into an action.
- Missing information is supplemented automatically by the operator with the help of his experience. For example, the instruction "Assemble a Flange" suffices to perform the described operation. The operator searches for the flange, places it on the assembly object and inserts the fasteners. Exact information about the position of the insertion holes is not necessary.

- The operator performs automatically a sensor controlled positioning operation. For example if he enters a screw into a hole, he uses several sensors. Course positioning is done with the aid of vision and fine positioning under the guidance of the touch sensors of his fingers. In case the thread of the screw does not engage with that of the hole, he instinctively takes correction action.
- Missing assembly element tasks may be automatically supplemented by the operator. For example the insertion of a screw implies that a screwdriver has to be used and that the fastening process has to be done according to a given sequence of operations.
- Fixturing needed during assembly may automatically be done by the operator without explicit instructions from the drawing.

It can readily be seen that many of the above mentioned tasks have to be programmed explicitly if a robot would perform the assembly. In order to obtain a fast and flexible programming system for robots, the following components should be provided:

- Installation of sensor for vision, force-torque, slip, proximity etc. into the robot.
- Control of the robot by a runtime system which is able to adapt itself to online changes during assembly.
- The compiler to translate the programmed work cycle should have a component which can automatically generate missing information.

These criteria imply that the robot should be controlled by a computer and that a higher programming language is available.

Fig. 3 shows desired features of programming languages for robots. In addition to the constructs of conventional languages, there should be several ones, specific for robots. For example, typical data types are vector, frame, rotation and transformation. It also should be possible to describe to the robot an effector trajectory and how to handle the synchronisation of the work of several arms. The robot must be able to operate the effector and the work tools under program control. In addition, there must be language constructs available which can handle signals to which the robot is capable to react.

2. The Structured Robot Language SRL

Based on the experience with AL (Assembly Language) and PASCAL the high level
robot programming language SRL was developed at the University of Karlsruhe.

2.1 Basic concepts

The language was designed as a powerful software tool to programming complex
robot tasks in the field of assembly using sensors. Therefore much emphasis is
on supporting a structured way of programming.

SRL is based on the above described frame concept. The geometric data types
VECTOR, rotation and frame are added and many powerful arithmetic operators
are included for geometrical calculation of robot positions and trajectories.
Especially for controlling two or more robots or for the evaluation of sensor
data during robot moves multitasking facilities are provided for parallel,
cyclic or time delayed execution of program parts. The task can be called with
actual parameters (like procedures) for passing data.

To overcome hardware dependence and to support structured and self-documenting
programming SRL includes language constructs to fulfill the goals mentioned
above. As a new facility SRL has an interface to a general world model at program run time. The world model can contain data about objects and their attributes, like workpieces, fixtures, robots, frames and trajectories.

Another fundament of SRL is the language PASCAL /4/. The data concept and file
management is taken from PASCAL because it gives the user a very flexible and
problem oriented data structure.

The goal of the development of SRL is the design of a language which can easily be learned and adapted to further developments and applications and also to
provide an interface between future planning modules and the "traditional"programming system. A planning module will be used to generate SRL statements
from a task (goal) oriented specification instead of explicit programming of
every action, see Fig.6 . Therefore SRL has to be well structured and universal and it has to include all features for robot programming and process control.

2.2 Concept of Data

The concept of data is based on PASCAL. The standard data types INTEGER, REAL,
BOOLEAN and CHAR are those from PASCAL. Also SRL includes the structured data
types ARRAY, RECORD and FILE of PASCAL. Furthermore the programmer can define

his own problem oriented data types as in PASCAL by enumeration types and sub-range type. Also pointers are included in SRL. As in PASCAL the programmer can write records and its components in any expression with respect to data types.

The geometric data types VECTOR, ROTATION and FRAME are introduced as prede-fined RECORDs. This guarantees an easy access to components of vectors, rota-tions or frames. A semaphor is used for the synchronisation and queueing with-in programs. A system flag SYSFLAG is needed for synchronisation between pro-grams. The programmer has no direct access to the data type SEMAPHOR or SYS-FLAG but he can use the statements SIGNAL and WAIT to handle them.

2.3 Specification of System Components

The language SRL includes a system declaration part for adopting programs to different sensors, robots and hardware facilities. With the help of the sys-tem specification the programmer can write programs which are more hardware independent, self documented and portable. If a program does not have a sys-tem specification the programmer has to use the predefined identifiers like ARM or CHANNEL.

Also digital and analog ports, absolute address registers, interrupts and flags to other programs can be specified. For having access to external data from a world model or teach-in the programmer can specify different files.

2.4 Program Flow Control and Structure

SRL includes the block concept of ALGOL. That means in SRL a block is not only a compound statement but it can also include a declaration part. A block may contain another block and the scope rules for local and global variables are the same as in ALGOL. In SRL a different statement is included for a syntacti-cally needed compound element to treat a sequence of statements as one.

The procedure concept allows passing parameters with call-by-value and call-by-address. If a procedure is declared with a type specification, the result of the procedure call will be a value of the specified data type (similar to the AL typed procedure or PASCAL functions). Additional to the normal declara-tions a procedure may contain declarations of procedure own variables.

These variables will hold their value from one procedure call to another but the variables are available in the procedure body only. This is helpful e.g. for programming counters.

Sections are tasks which can be executed parallel to other sections.
SRL contains the traditional control structures of PASCAL:

- IF THEN ... ELSE ... END_IF

- CASE OF ... END_CASE

- FOR STEP ... TO DO ... END_FOR

- WHILE DO ... END_WHILE

- REPEAT UNTIL ... END_REPEAT

All statements are closed by an END_ statement . To avoid undefined situations
the CASE-statement includes an OTHERS-exit.
To abort the execution of a program branch, a loop, a section, parallel parts
or a program SRL includes an EXIT statement. Program control returns after the
EXIT statement to the outer control level of the program.

2.5 Move Statements

To distinguish between different types of interpolation SRL includes several
move statements. It depends on the implementation and the robot control which
of the move statements are available. The specification of position and orientation of the tool center point (TCP) is based on the frame concept. Reactions
on sensor conditions or data are handled by the WHEN or ALWAYS WHEN statement.

PTPMOVE   Move without any synchronisation between the robot axis.
                Each axis is moved with maximum acceleration and speed. No
                general specifications are allowed.

SYNMOVE   Linear interpolation in robot joint coordinates, i.e. all
                axes will be synchronized. General specifications possible.

SMOVE   Movement on a straight line by linear cartesian interpola-
                tion. General specifications allowed.

LANEMOVE   Trajectory calculation by polynoms, similar to the MOVE-
                statement of AL. General specifications allowed.

CIRCLEMOVE   Movement along circular segment. Specifications: center
                point, angular displacement, velocity or duration, fine/
                rough interpolation and positioning.

VIAMOVE   Move to a via frame without stopping at the via frame. The
                interpreter expects a next move statement for continuing
                the move. Only special specifications of velocity and dura-
                tion are allowed.

MOVE   Move statements with interpreter or controller dependent parameter specifications. This statement can be used for all future types of interpolation if the controller includes the control modules.

DRIVE   Movement of one or more robot axes. Specifications: velocity or duration, force, fine/rough positioning.

The general specifications for a trajectory controlled by the programmer are:

- velocity (VELOCITY)
- duration (DURATION)
- acceleration (ACCELERATION)
- constant/variable orientation during the move (CONSTORIENT, VARORIENT)
- approach/departure points (APPROACH/DEPARTURE)
- frames between start and end frame (VIAFRAMES)
- force (FORCE)
- wobble movement (WOBBLE)
- robot arm posture (POSTURE)

2.6 Access to Data Defined by Teach-in

With the help of the interactive component the programmer can define frames and technological parameters by teach-in. These data are stored in a framefile. The main statements for the predefined type FRAMEFILE are:

FRAME INITIALIZE (framefile)

                All frames of the actual block in the program which have the same name as the frames in the internal framefile get their values from the framefile.

WRITE FRAMELIST (framefile , frames)

                Adds the specified frames to the framefile or overwrites their values in the framefile.

2.7 Access to Data of a World Model

Many data describing the robots, machines and objects is stored in a data base. This world model is not only important during planning and programming but also at runtime. Therefore SRL includes an interface to a world model at program rune time.

The programmer can ask for the existence of an object described in the data base by using the system function:

EXISTENCE (objectname)

It returns a boolean value: Reading the value of an attribute of an object:

variable := ATTRIBUTE name OF objectname ;

The program can also change a value in the world model:

ATTRIBUTE name OF objectname := expression ;

The affixments of AL are done in the world model by the statements

AFFIX object1 TO object2 ; and

UNFIX object1 TO object2 ;

The effect is exactly the same as in AL.

If the user wishes to have stored the actual state of the world by sensors he can write

UPDATE ATTRIBUTE name OF object BY
sensorname EVERY expression MS
WITH expression ;

2.8 Interface to Robot Control

After the programmer has written his SRL-program it will be compiled and translated into IRDATA-code. IRDATA is defined by the working committee of VDI (German Engineers Association) with a main contribution of the author. After transferring the user program as an IRDATA-text to the robot control it will be executed by an IRDATA-interpreter. Therefore SRL can be used for any robot control which is supported by an IRDATA-interpreter.

2.9 Programming Components for SRL

The programming environment (Fig. 4) includes the major parts:

- SRL-compiler
- IRDATA-interpreter
- Frame-editor
- Interactive component
- World model (optional)
- Simulator (optional)
- Symbolic debugger (optional)

The SRL -compiler reads in a SRL-program text. checks the syntax and generates
similar to the P-code in PASCAL an IRDATA code, which is executed by the IRDATA
interpreter. The frame-editor and the interactive component are used to specify
frames offline or online by teach-in. The frames are stored in a frame list and
added to the program at runtime.

3. Robot Data Base   R O D A B A S

Using SRL as a high level explicit programming language the programmer has to
define every frame of the robot moves and every action. This is time consuming
and can lead to errors if the programmer has not in mind all detailed information. During program execution the robot control should be able to react on unforeseen event or assembly errors. Solving these problems can be done by using
methods of Artificial Intelligence (AI) to describe objects, structures and
action sequences. Based on a special relational data base a knowledge base was
implemented which includes the description of the objects to be handled, the
working sphere of the robot, the magazines, the transport systems, obstacles
and other features of the robot environment.

When RODABAS is applied to systems for integrated manufacturing control there
are data of other fields like geometrical data or assembly data sheets from a
CAD data base. With access to this background knowledge a planning module can
generate the needed robot moves and trajectories depending on the description
of objects and robot environment. During runtime with the help of sensor modules an updating can be done to provide data for error monitoring and error
correcting, see Fig. 6.

3.1

The representation of any feature and structure requires a scheme which allows a description in such a way, that it can be stored and managed by computers. The FRAME-concept of AI is based on the basic abstract features, ob-

jects, attributes and attribute values. Object means a unit which can be identified and an attribute can be related to an object and can have different values.

There is also a classification of objects with equal attributes into different types or classes. The user is able to describe hierarchically structured classes or networks between them. The basic structure (object, attribute, value) is transformed directly into a relational data base model. The class of objects of the same type corresponds to a relation which is represented by a table. Every row stands for an attribute, the table itself includes the values of the attributes.

3.2 Implementation

For defining or deleting relations (object classes),new objects and attributes, there is at any time a realization necessary which combines flexibility with the simple structure of the table. The RODABAS-implementation guarantees this requirement by dividing the tables in the direction of rows and columns into blocks of fixed length, which are linked together. By adding or deleting of block chains there is an extension or reduction of objects or attributes possible, see Fig. 7. The objects and attributes are identified by names which allow a fast access using hash-tables to calculate the table position.

Every attribute contains an attribute description to realize values of different types and lengths. The attributes descriptor includes the data type of the value, its maximum size, the range value and default value. The RODABAS-implementation allows more information, e.g. the unit of the value if it represents a physical dimension. From the frame concept is derived an object description including marker for an representative object for a type or class with common attributes or a pointer to a type templet.

3.3 Data Base Functions

The data base functions can be divided into:

- structure manipulation (generate and delete relations, objects or attributes)
- data manipulation (input, correction or delete values)
- evaluation (output of single values or value arrays)
- help and information

The data access is supported by a comfortable cursor module which allows a systematic search under conditions in the relational tables. The dialog interface is menu -oriented and easily used. There is a help function for information about commands and their parameters.

3.4 Open-Ended-Design of the World Model

Based on the above described relational data base a "kernel" of a world model
for assembly operations was specified. A scheme of description was predefined
and the user can apply it to describe his robot, sensors, move sequences and
other features. Fig. 8 shows the predefined structure of a move in RODABAS,
where the user has to give only the values and references. This structure of
the object description should not be changed normally but if it is necessary
to extend it for special purposes, e.g. to add information about the temperature of an object, the structure can be changed easily.

The result is on one side a finished tool for solving standard problems of the
user and on the other side the model can be adapted or extended to any new
application very flexibly.

References:

Blume, C., Dillmann, R.:
Frei programmierbare Manipulatoren (Free Programmable Manipulators)
Vogel-Verlag, Würzburg, Germany, 1981

Blume, C., Jakob, W.:
Programmiersprachen für Industrieroboter (Programming Languages for Industrial
Robots, Vogel-Verlag, Würzburg, Germany, 1983 (will be published in English by
Springer, 1984)

Blume, C., Jakob, W.:
Design of the Structured Robot Language (SRL), Proc. of Advanced Software for
Robotics, Liege, Belgium, 4-6 May, 1983

Jensen, K. and Wirth, N.:
PASCAL-User Manual and Report, Springer 1978

Rembold, U. and Blume. C.:
Programming Languages and Systems for Assembly Robots, Computers in Mechanical Engineering, January 1984

D'Souza, C., Zühlke, D., Blume, C.:
Aspects to Achieve Standardized Programming Interfaces for Industrial Robots,
Proc. of 13th International Symposium on Industrial Robots, Chicago, 1983

Müller, E. and Pods, R.:
Entwurf und Implementierung einer Umweltmodell-Datenbank für die Roboterprogrammierung, Diplom Thesis, University of Karlsruhe, 1982

Blume, C.:
Wissensbasis für die zukünftige Roboterprogrammierung (Knowledge Base for
the Future Robot Programming), ELEKTRONIK 16, August 1983

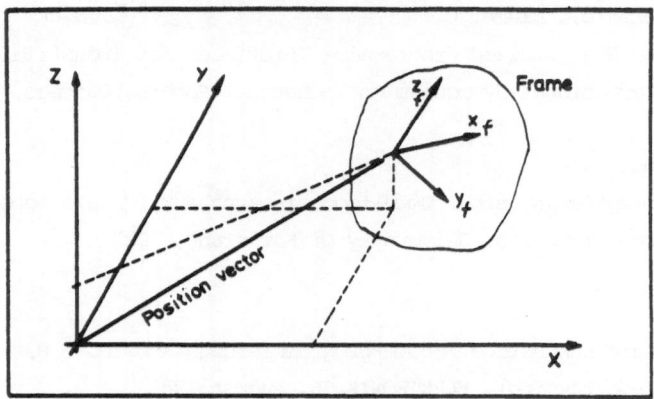

Fig 1: Geometric position af a frame

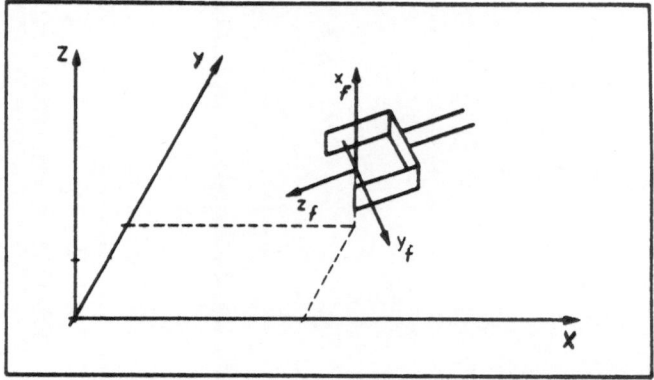

Fig 2: Geometric position and orientation of a gripper

Teach-in programming
Control structure
Subroutines
Nested loops
Data types
Comments
Trajectory calculation
Effector commands
Tool commands
Parallel operation
Process peripherals
Force-torque sensors
Touch sensors
Approach sensors
Vision systems

Fig 3: Features of programming
languages for Robotics

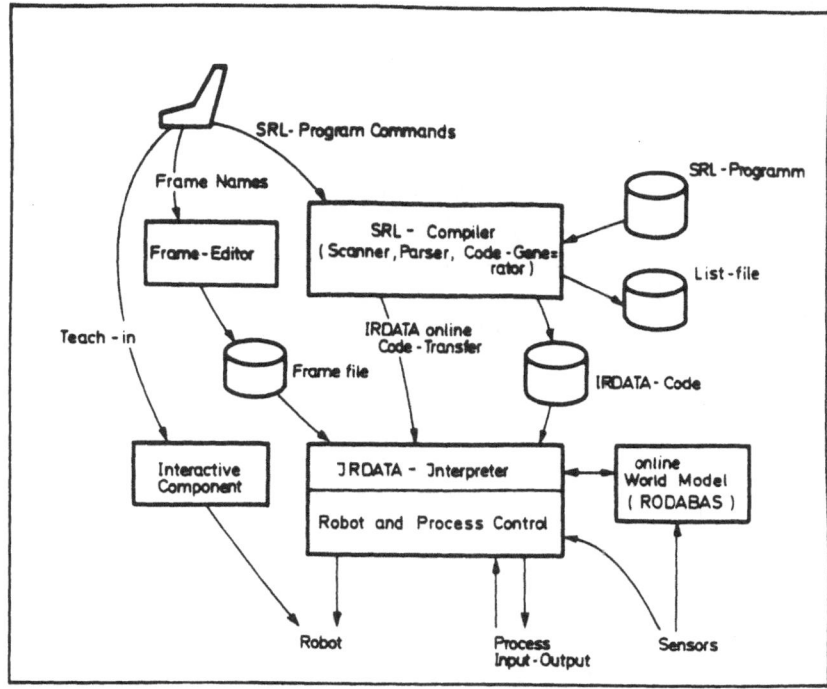

Fig 4: SRL - Programming - System

```
(*#############################################################################)
(*                                                                            *)
(*  Example for an explicite programmed harmonic movement by use of recursion. *)
(*                                                                            *)
(*#############################################################################)

SYSTEM_SPECIFICATION
ROBOT     puma      = ARM (1);
DATABASE surrounding = FRAMEFILE ('PUMASTATION.FRF');
END SYSTEM_SPECIFICATION;

PROGRAM harmony;
CONST vmax    = 80;                       (* Maximum speed of the robot  *)
      delta_t = 0.25;                     (* Timeinterval                *)

VAR    startposition, endposition : FRAME; (* Values from the framefile   *)
       halfway, direction : VECTOR;
       acceleration : REAL;

PROCEDURE harmonic_move (range: VECTOR; actual_velocity: REAL);
VAR actual_acc : REAL;          (* actual acceleration *)
BEGIN_PROCEDURE (* harmonic_move *)
  actual_acc := acceleration;   (* Need not be a constant during the movement. *)
                                (* E.g. a function of the passed way.          *)
  range := range + (0.5*delta_t*delta_t + actual_velocity*delta_t) * direction;
  actual_velocity := actual_velocity + actual_acc * delta_t;
  IF (|range| < |halfway|) AND (actual_velocity <= vmax) THEN
  BEGIN
    VIAMOVE puma TO startposition + range WHERE VELOCITY = actual_velocity;
    harmonic_move (range, actual_velocity);
    VIAMOVE puma TO endposition - range WHERE VELOCITY = actual_velocity;
  END;
END_PROCEDURE (* harmonic_move *);

BEGIN_PROGRAM
  RESET FRAMELIST (surrounding);        (* Initialising the frames            *)
  FRAME INITIALIZE (surrounding);       (* startposition and endposition      *)
  CLOSE (surrounding);                  (* and deleting the internal framelist *)
  halfway := (endposition.transl - startposition.transl) / 2; (* Initialising *)
  direction := halfway / |halfway|;     (* basic data                         *)
  WRITELN (ttw, 'Please, enter the acceleration');  (* for the               *)
  READLN (ttw, acceleration);           (* movement                           *)
  harmonic_move (nilvector, 0);         (* Starting the recursion             *)
  MOVE puma TO endposition;             (* Final MOVE to end the VIAMOVE's    *)
END_PROGRAM.
```

Fig 5: SRL - example

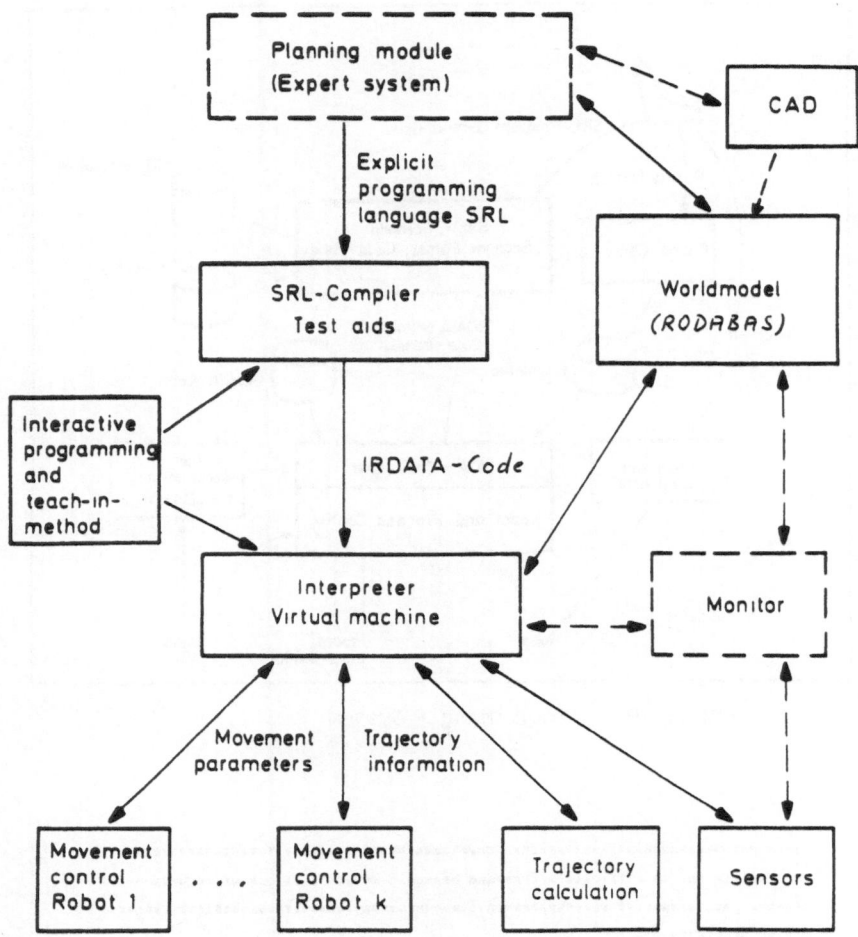

Fig 6: Hierachical programming structure for
        industrial robots

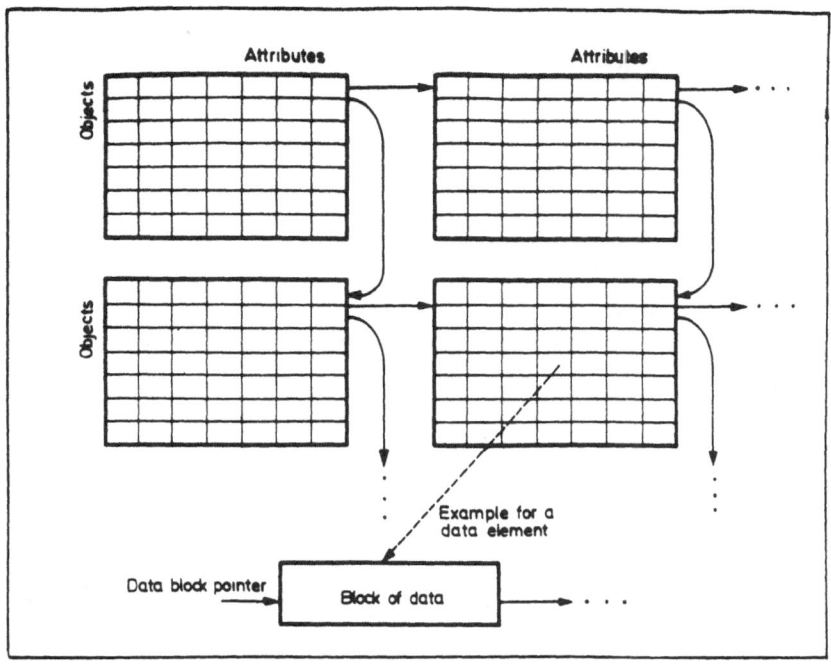

Fig 7: Structure of the RODABAS - implementation of the relational table

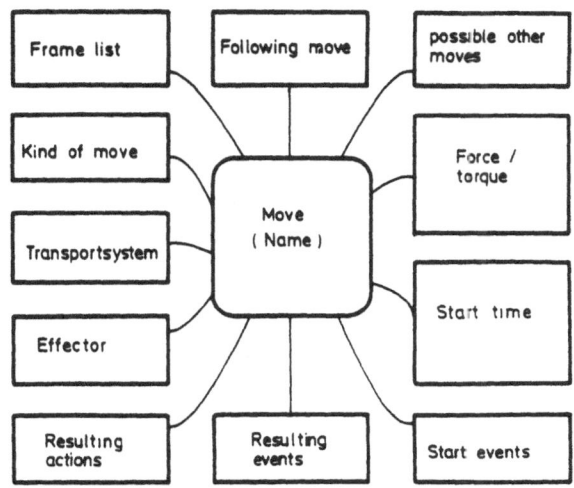

Fig 8: Predefined structure of description of a move in RODABAS

Robot Architecture for the Integration of Robots into Manufacturing Cells

R. DILLMANN

Institut für Informatik III
Research Group Robotics
Director: Prof. Dr.-Ing. U. Rembold
University of Karlsruhe
7500 Karlsruhe, West Germany

Abstract

Actually the number of flexible manufacturing systems incorporating
industrial robots is relatively small. The integration of robots into manufacturing cells has been realised for different applications, but a general
method is still missing. One reason is that software and hardware linkages
between machine tools, robots, sensors and peripherals are not standardized
because of their different types of data. Data generation, presentation and
manipulation in the field of robotics is till today technology dependant
and inconsistant. A structured consistant data concept and a structured
system architecture is the key for a CAD/CAM/Robot linkage. This paper dis-
cusses a hierarchical robot architecture based on a hierarchy of task decomposition operators which operate on virtual robot data.

1. Introduction

There is a increasing demand in industry for manufacturing cells consisting
of multiple robots, machine tools and peripherals which are integrated into
a functional unit. Different cell types like machining cells, welding cells
or assembly cells are to be configured depending to the application. A machining cell for example can consist of a milling work station combined
with a inspection work station and a material handling work station. The
workstations are functional combinations of robots, machine tools, part
buffers, peripherals etc. Vision system and their combination with other
sensors are basic elements in the workstations which aid high pertormance
finishing, complex assemblies and piece part inspection for quality control.
Under the context of CIM (computer integrated manufacturing), manufacturing
cells are planned and programmed in an environment which is characterized
by an unique data model of the factory with a strong integrity of data and
a defined data transaction management. For this purpose, the integration
of CAD/CAM and robotics is expected in this decade. There are worldwide
national and international projects pursued in which robots are to be inte-

grated as a part of a complex production system. One approach is to treat robots like a NC machine which yields to DNC concepts using APT like languages for programming. The use of post processors allows interfacing of different robot types. For sensor guided robot applications with intelligent control capabilities which are needed for assembly operation a more flexible control concept is necessary. Albus /1/ proposes a system in which robots are integrated via a tree type control structure into a manufacturing cell. The CAM-I software project /2/ claims a manufacturing system model which defines a linkage between CAD/CAM, programming and simulation systems and the robot control. Within the framework of the European ESPRIT project, design rules for the integration of industrial robots into CIM systems are under study.

The data presentation of a robot in a manufacturing cell, the data processing and the data flow to control robots within a cell is called robot architecture. From this architecture a computer soft and hardware concept for robot controls and an adequate operational principle can be derived. This computer architecture fullfills the requirements and integrational aspects for robots in a manufacturing environment. The robot architecture of the Karlsruhe Robot Research Group is emphasizing intelligent multiarm control, the integration of multisensor systems, peripherals and tool machines. A system architecture is proposed which is hierarchically structured using a distributed polyprocessor system /3/. Robot dependant and robot independant control levels can be identified according to the control task. The use of standard interfaces /4/ between the hierarchical ordered control levels allows the integration of each robot type if appropriate interpreters are available. Interfaces to higher off-line control levels like textual programming system, simulation systems and data bases increase the exploitation of the robot manufacturing potential. In the following the robot architecture and the control computer concept is presented and discussed.

2. Robot Architecture

The internal presentation of a robot in a manufacturing cell is assumed to be hierarchical. The control system of the cell is composed of a multitude of individual functions which assure the performance of the manufacturing task, the safety and emergency aspects, the linkages to other manufacturing cells and the interconnection to higher manufacturing control levels. The problem of organizing and coordinating these particular control functions as well as the problem of defining a logic control structure implies the

definition of five hierarchically ordered system control levels. The five
distinguished tiers are related to (see Fig. 1a,b)

- axis, end-effector and primitive sensor controls
- arm- and local peripheral control
- robot, global sensor and peripheral control
- cell control
- multiple cell control (shop-floor).

The hierarchical tiers are composed of functional modules (soft and hardware) which perform specific control functions and tasks. The logic combination of these functional modules on each specific control level yields a
horizontal control structure. This structure contains manipulator end effector and sensor controls as well as emergency treatment and human safety
control structures. The vertical data flow between the specific control levels is represented by control commands, sensor data and emergency detecting
messages. A basic description of hierarchical organised robot control
structures is discussed in /5,6/. The idea is to use a series of hierarchical ordered task decomposition and task sequencing operators which have as
input on the highest level a manufacturing task description command. This
command is to be decomposed via the hierarchical ordered operators into a
sequence of basic control functions and primitives, which have to be executed in real time. The system is characterized by distribution of intelligence. Thus, sophisticated intelligent control structures can be organized
in a hierarchically decreasing order of abstraction and intelligence and
in an increasing order of precision and sampling rate. The use of intelligent modular subsystems for building hierarchical organized manufacturing
cell controls has the following benefits.

- Reduction of organisational efforts
- Extendability of system
- Flexible control structure
- Vertical and horizontal data flow between the subsystem is reduced
- Reliability through modularisation
- Reduced complexity of software
- Reduced costs for control software development
- Realisation of subsystems in software or hardware
- Structures enables CAD/CAM/robot linkages
- Alternative design of simple, intelligent sensor driven or highly
  sophisticated controls possible
- Independant development of robot, NC and peripheral subsystems
- Use of standardized linkages between different control levels.

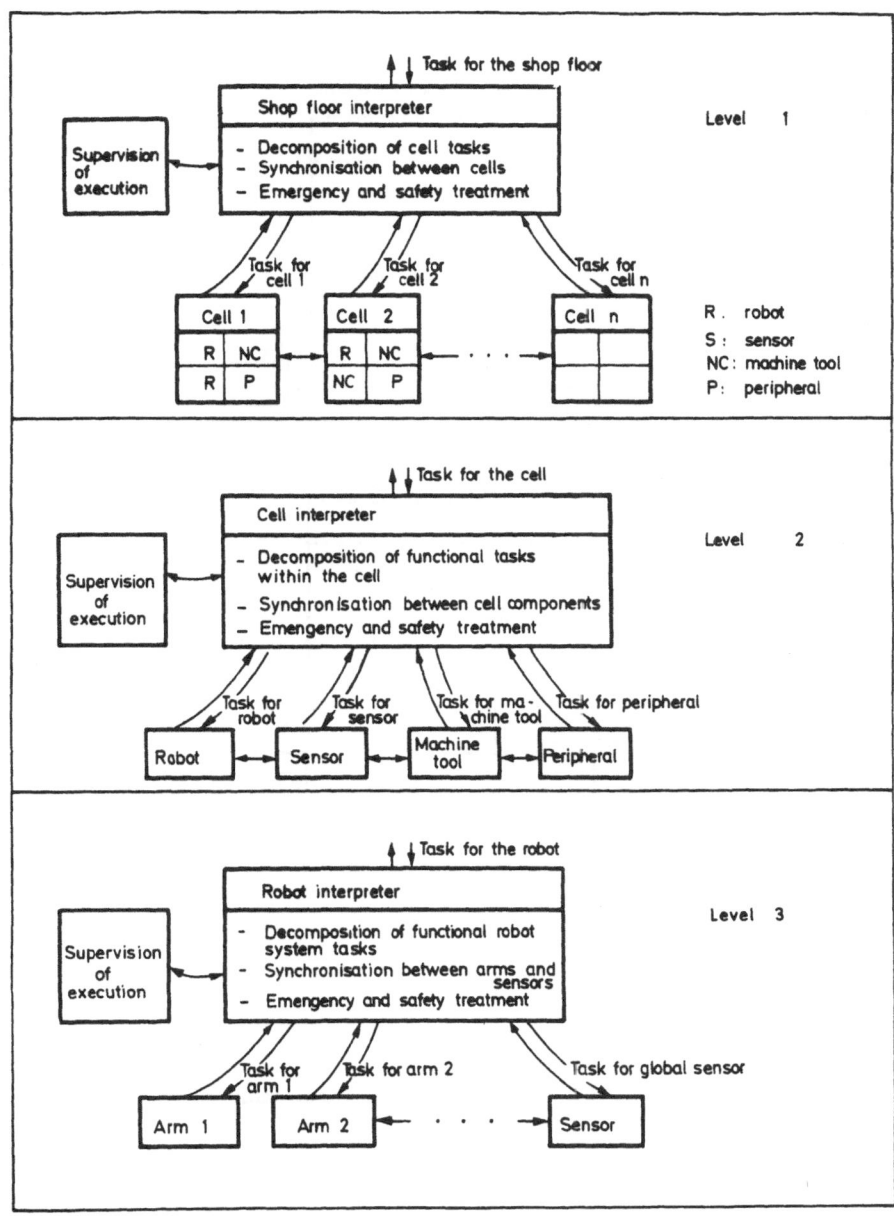

Figure 1a: Basic control structure for a robot oriented manufacturing task (robot independant)

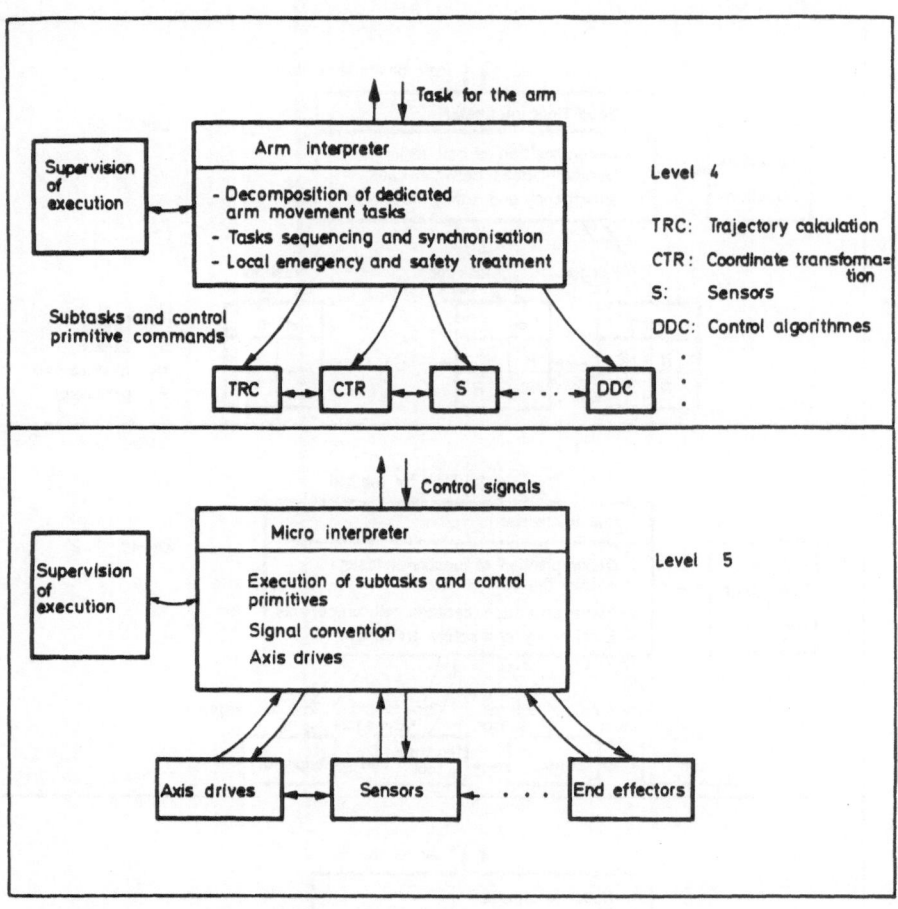

<u>Figure 1b</u>:   Basic control structure for a robot manufacturing task
          (robot dependant)

The architecture of the robot control system is defined by
- The internal robot data presentation
- The task decomposition operators
- The horizontal and vertical flow and
- The operating principle of the modular subsystems.

The internal robot data presentation is related to data objects which are to be processed by control level specific operators. Such operators can be task decomposition operators, sensor and sensor monitoring operators, world model operators, conflict-analysis and decision operators etc. The mechanism of the task decomposition defines the relation between two control levels and the relation between sensor information, action and expected results. The horizontal and vertical data flow defines the software and hardware linkages between subsystems and between diferent control levels. The operating principle of the subsystems defines how to start, to interrupt and to finish a task.

This robot architecture contains information, which is essential for the design of computer controls, robot programming languages and robot data bases.

3. Internal Robot Data Presentation

Within the five tiers defined for the control of manufacturing cells the view of industrial robots is quite different. Seen from higher levels the robot presentation is closely related to the manufacturing application. The robot appears on the shop floor level as a component of the different manufacturing cells. The manufacturing cells are designed for a specific task and defined by their functionality. The shop floor is controlled by an interpreter which decomposes the cell tasks according to the shop-floor manufacturing program. This includes the synchronisation between cells, the material flow and the emergency treatment. The robots do not appear explicitly on this level. Each cell is defined by it's functionality, it's task capacity, it's elementary operations, performance, etc. To solve the problem of cell task decomposition, the interpreter has to operate on data related to a priori and actual process knowledge and virtual cell descriptions. Output of the shop floor interpreter is a sequence of instructions for each cell, Fig. 2. The manufacturing cells are configured by different combinations of robots, machine tools, peripherals and sensors. The cell interpreter controls the operation of all functional devices within the cell. The

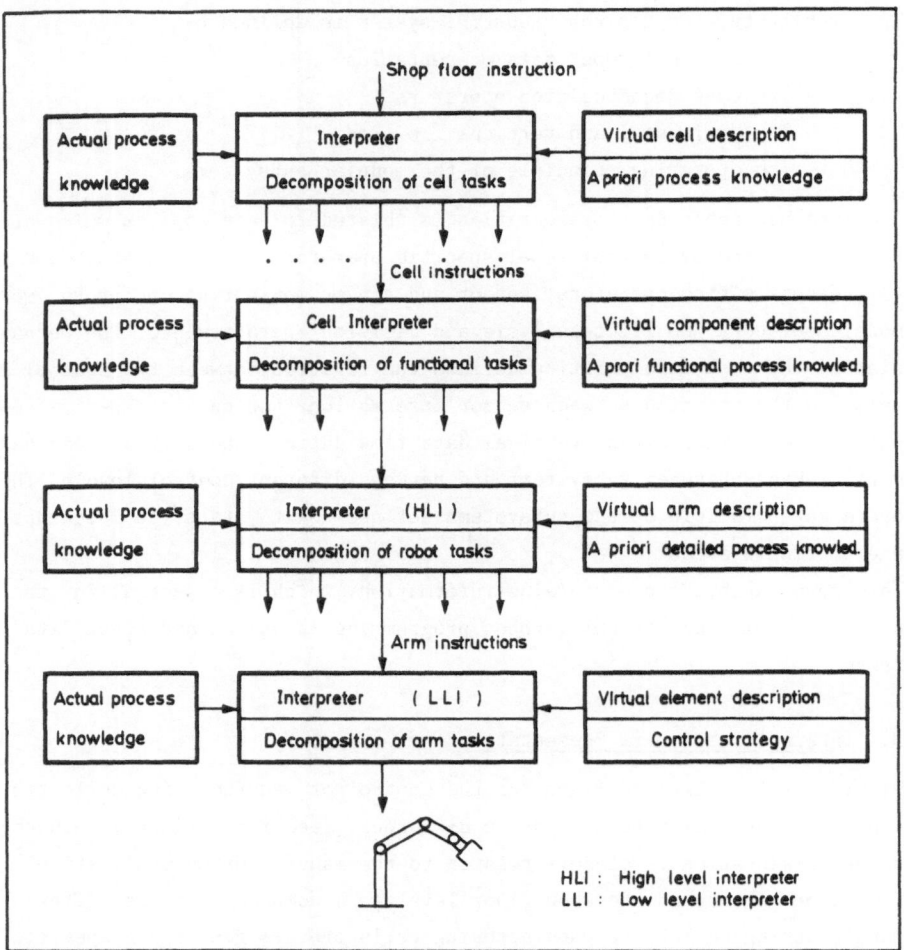

Figure 2:   Robot task decomposition within a manufacturing cell

input instruction is decomposed into device specific tasks (robot, sensor, NC, peripheral, etc.). Emergency treatment and safety tasks influence the task sequences using actual and a priori process knowledge. Robots are app arent on this level by their functionality, elementary operations, process capabilities and performance in terms of virtual description. The output of the cell interpreter is a task sequence to be executed by the robot, the machine tool, the peripherals, etc. These tasks are to be executed in parallel and/or synchronized. The mechanism of hierarchical task decomposition is continued via the robot and arm interpreter. The principle is to break down a large abstract problem into a set of smaller, more detailed subproblems. At each level of decomposition, details and facts are filled in to make the next lower subtask series more concrete. At the lowest level of the hierarchy exists an ordered number of independantly solvable problems and functions. The details and facts which are needed for task-decomposition have to be presented internally in the control architecture. There is a large number of required data and descriptions which are related to the manufacturing cell, to the tool machines, robots and sensors. This data is to be presented in form of functional, kinematic and dynamic models. Grippers, tools, conveyors, links of the robot but also trajectories, positioning tables and uncertainty information are some examples of the required data. In addition geometric and kinematic information about the geometry of application, the resolution of sensor and image processing is needed. The timing of processing and the type of branching according to the sensor information has to be defined. Sensor models for the sensor hypothesis which allows the decomposition of a sensor plan must internally presented. From the a priori uncertainty information available sensors (proximity, touch, force wrist, 2D nd 3D visual, compliance, etc. constraints, restrictions and safety requirements are also to be taken into consideration for task decomposition. Depending on the application, the presentation of process knowledge which might be a catalog of possible actions, reactions and their relationship to preconditions (e.g. sensor pattern) and their effect to the overall system is needed. All this information is taken together to decompose the tasks. To structure this amount of information in a sense of modularity specific processors which operate on the required data types can be defined. This yields to specific processors like

- A sensor monitor
- A conflict analysis processor

- A decision processor
- A geometric world processor
- A trajectory management processor
- Processors for axis, end-effector and sensor control.

Each of these processors operate on task specific knowledge and produce inferences necessary for task decomposition and branching in the decision tree. To limit the amount of data, on higher levels only virtual concentrated information about the lower levels is to be presented, Fig. 3. In other words, instead of an extensive list of details and facts a concentrated abstract description (e.g. functional) can be used. This data can be stored in a data base or for the on-line process in the memory of the computer architecture. Figure 4 shows the principle of data acquisition and presentation within the control hierarchy.

4. Task Decomposition and -Execution

Task decomposition in herarchically organized control structures has been outlined by Nilsson /7/ very early. A hierarchical robot control concept consisting of task decomposition, sensor and predictive world data operators was studied by Albus et. al /1, 5/. Starting from a manufacturing cell task description, a hierarchy of task decomposers operate on problems with decreasing degree of abstraction with the goal to produce a sequence of states or actions which lead from the initial state to the goal state in an optimal sense. Constraints and uncertainties are taken into consideration to reach this goal. The generation of these state sequences requires the simultaneous qualitative and quantitative presentation of actual world and robot data. The initial system states and the possible actions and reactions supported from the catalog of available control functions are to be evaluated by a decision operator which has to find the optimal solution in the search space. Decision making is assumed as a graph searching problem on decision trees whose nodes are possible task states and whose arcs are subsequent actions or subtask sequences.

After task decomposition the reaching of the goal state within a defined time interval is supervised. If constraints or emergencies are detected a new decomposition cycle will be initiated.

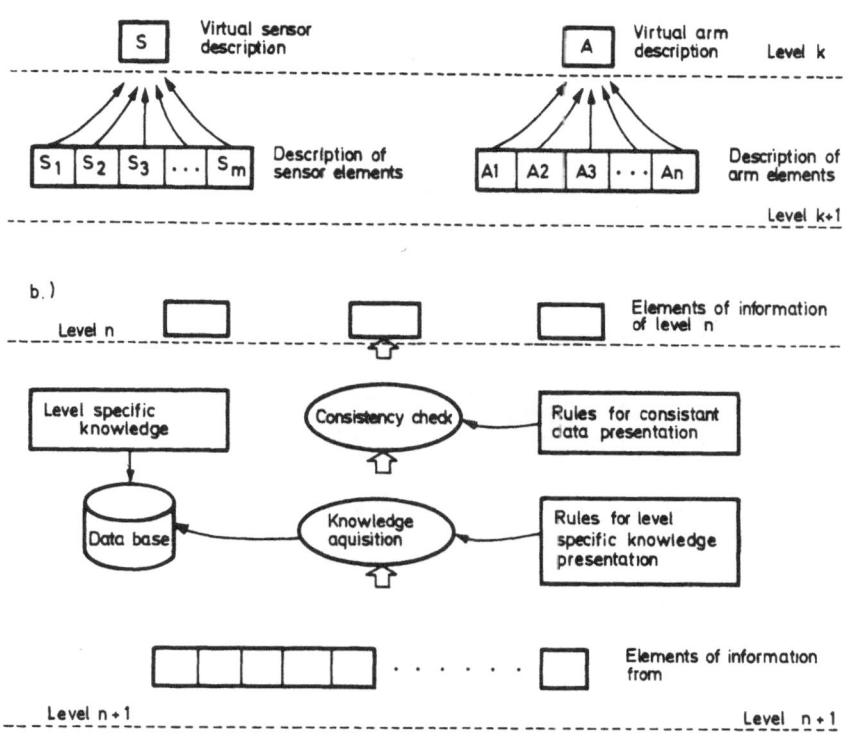

Figure 3: Priciple of data structuring for robot data presentation

      a. Concentration of data in terms of virtual description

      b. Diagram for consistant knowledge acquisition to virtualise detailed information from a lower to a higher control tier

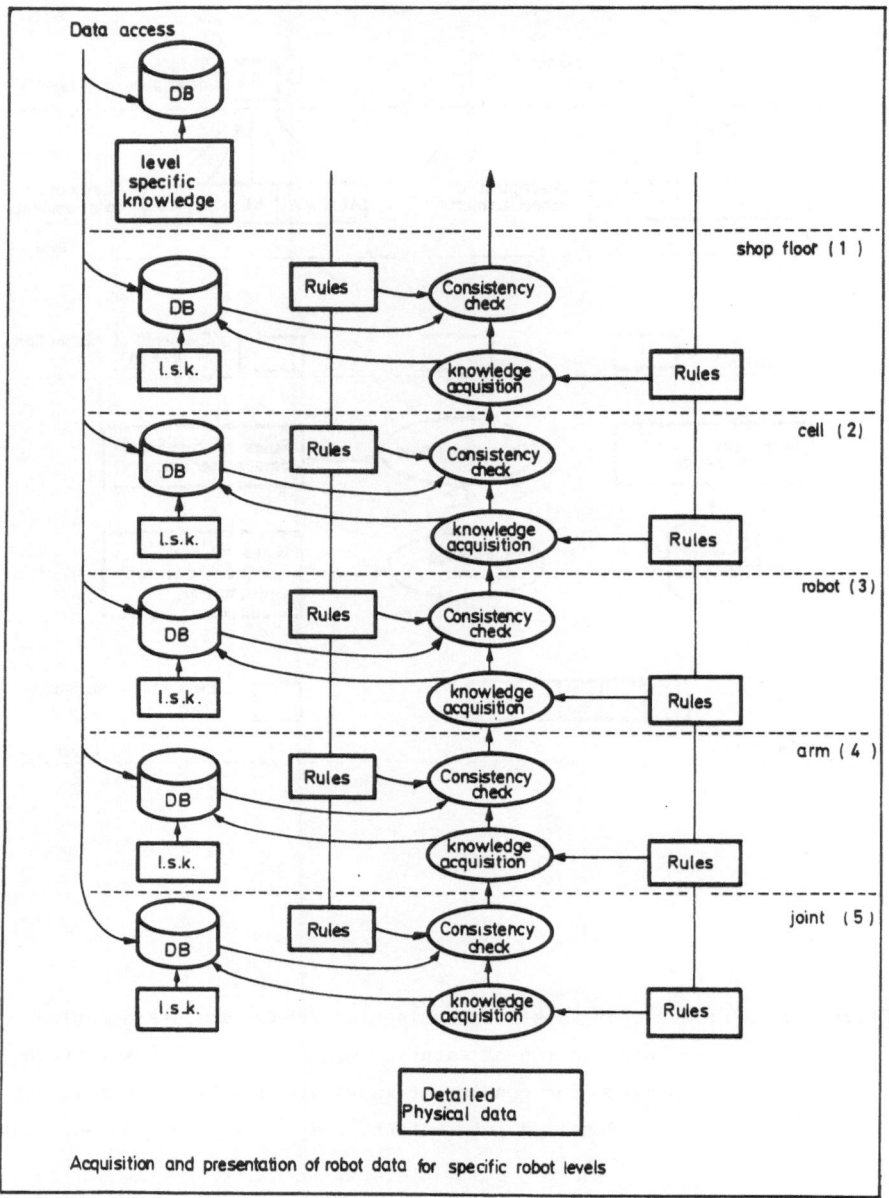

Figure 4: Acquisition and Presentation of data for specific control
tiers

5. Data Flow and Computational Concept

The horizontal and vertical data flow within the robot architecture is
illustrated in figure 5. Control instructions are transmitted from higher
to lower tiers for decomposing task sequences. From lower levels sensor
data, errors, emergencies, status etc. are transmitted to support task
decomposition on higher levels. The horizontal data flow is located
between the task-decomposition operators,sensory processing operators
and world-model operators. Each task decomposition operator is supported with extracted sensor data from the sensory processing modules.
The degree of abstraction of the extracted sensory data increases with
higher control levels. The sensory processing operators are provided with
a stream of task related state expectations and predictions including uncertainty information from the world-model processor at that level.Each
module receives inputs describing the task sequences, action plans and
hypotheses generated as output of the decomposition module. Emergency
detection and analysis, Fig.6,is performed by the sensory processing
module together with the world model module. Decision making and super-
vision of the task execution is to be done by the decomposition module.

   Seen from the programming perspective, the hierarchical structure of
the cell control is a structured top-down program applied to robots,
machine tools, sensors and peripherals. Each block represents a sequence
of instructions and a set of structured branches.
In terms of system architecture the computing modules can be assumed as
processes or finite state automatas. With this terminology each module
(operator) is an autonomous process which operates on an input data
stream to transform it into an output data stream. All operators can
be implemented on a microcomputer which transforms and exchanges date
and messages. This structured approach allows modularity in hard- and
software. Suitable computer architectures for such a modular system are
distributed systems and extendable symmetric polyprocessors /3/. Linkages to CAD/CAM systems are formed by the internal robot and world models,
which are presented in terms of frames or in terms of virtual robot and
process description.

5. Conclusion

A hierarchical modular ordered robot architecture has been outlined. The
architecture includes the functions of the manufacturing cell,the robot,
the arm and the axis in a logic functional internal presentation. Opera-

184

tors for task decomposition, sensory processing and world model manipulation support a data stream, which is necessary for the execution of intelligent manipulation tasks. Such a functional architecture forms a
linkage to the application problem formulation. Integration of robots
into manufacturing cells means to configure the robot architecture
with appropriate machine tool and peripheral control tiers. Efforts
are made by the Karlsruhe Robot Research Group to develop a polyprocessor architecture with the purpose to implement the outlined robot
architecture. Linkages to high level robot programming systems like
AL and SRL , to a robot data base (RODABAS) to a simulator and the use
of IRDATA are basic system features.

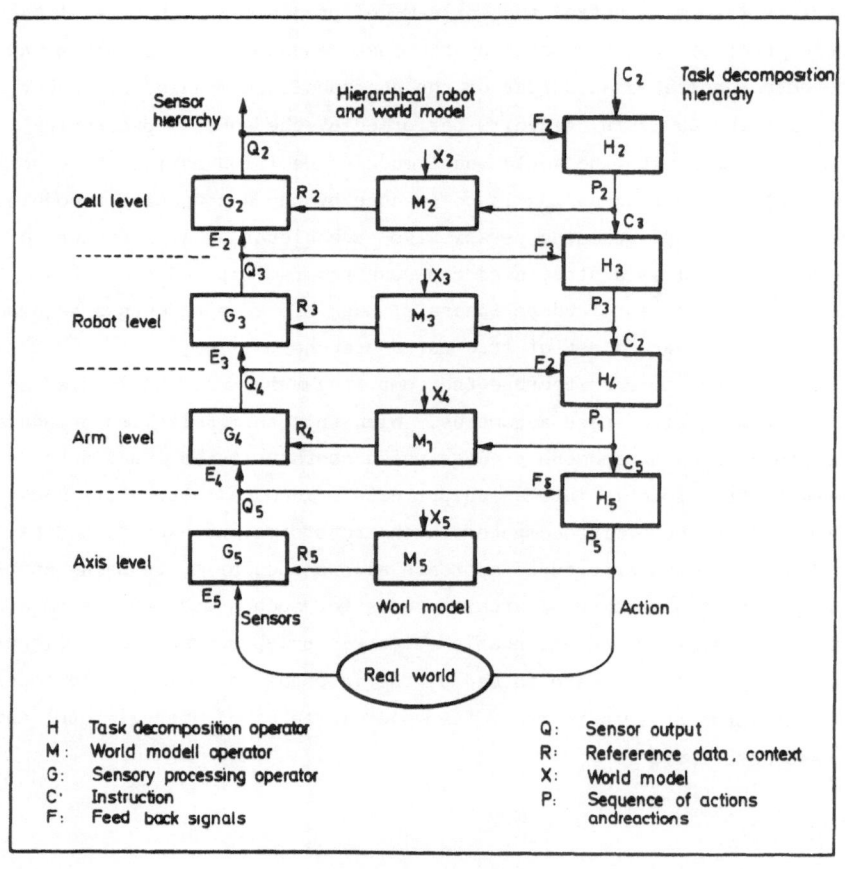

Figure 5:   Principle data flow in the robot architecture

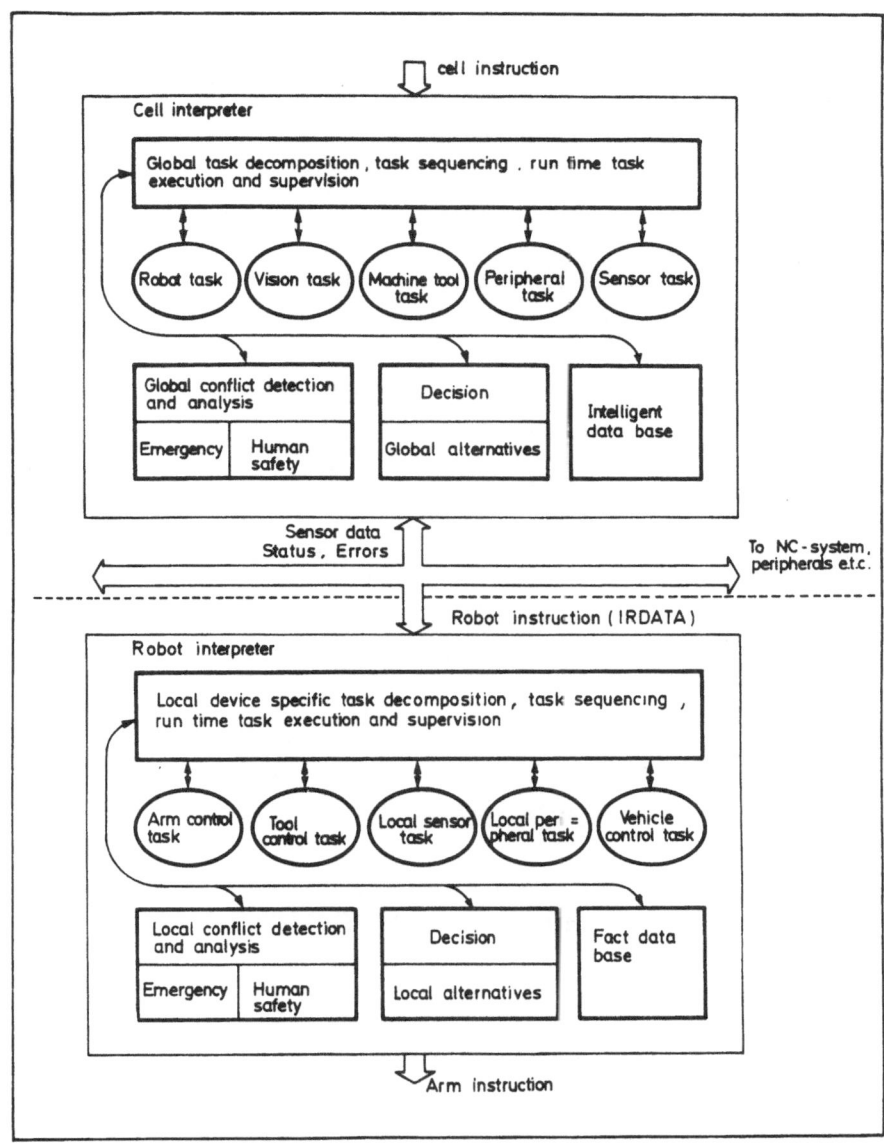

<u>Figure 6:</u>   Control structure of the cell and the robot for task execution

186

References

/1/   Albus, J. S. et al: "Hierarchical Control for Robots in an Automated
       Factory" Proceedings of the 13th ISIR/Robot 7, April 17-21, 1983,
       Chicago, Illinois

/2/   CAM - I Robotics Software Project 1983: "RSP Working Group Final
       Report, Outline of Functional Specification for Prototype Software"

/3/   Kordecki, C., Dillmann, R.: " Conceptual Design of Adaptive Multi-
       robot Control",Proceedings of the 1984 ASME International Computer
       in Engineering Conference,Aug. 12-16,1984,Las Vegas

/4/   Blume, Ch., D'Souza,Ch., Zühlke,D .: "Aspects to Achieve Standardi-
       zed Programming Interfaces for Industrial Robots", Proceedings of
       the 13th ISIR/Robot 7, April 17-21, 1983, Chicago Illinois

/5/   Albus, J. S. et al: "Theory and Practice of Hierarchical Control",
       23rd IEEE, Sept. 13.-17. 1981, Washington

/6/   Saridis, G. N., Stephanon, H. E.: "Hierarchically Intelligent Control
       of a Bionic Arm", Proceedings of the Conference on Decision and
       Control, December 1975, Houston, Texas

/7/   Nilsson, N. J.: "A Hierarchical Robot Planning and Execution System",
       Artificial Intelligence Center, TN-76, SRI-International, 1973

The Development of a European Benchmark for the Comparison of Assembly Robot Programming Systems

K. COLLINS, A.J. PALMER, K. RATHMILL
Cranfield Robotics and Automation Group (CRAG),
Cranfield Institute of Technology,
Bedfordshire, England

1.0   Introduction

Many robot manufacturers claim that their machines are suitable
for assembly operations, but before embarking on an automated
assembly project, a user would be well advised to look beyond
the mechanical performance of the rival machines and to look
closely at the software features offered.

It is not intended in this paper to compare the mechanical features of any robots but to examine the software and control features that apply to the assembly process.

In order that dissimilar devices could be compared in an unbiased
manner, a benchmark specimen was designed which incorporated the
most frequently encountered assembly problems. From this Benchmark a number of robot control features emerged which were either
essential or helpful to the assembly process.

It was then possible to look at several types of robot control
systems and assess the merits and demerits of each. Other factors were also taken into account such as the time taken and the
degree of programming expertise required to create the control
program.

1.1   Background

There has been increasing activity in the comparison of alternative robot languages within many industrial nations, not least
the United States of America (2)(3). However with growing interest in the possibility of commercially led defacto standards
being established, (4)(5) there is clearly a pressing need for

a meaningful, industrially oriented basis for establishing the merits of alternative programming systems.

2.0   Establishing a Benchmark

Arising directly from discussions held in the Robotics Europe (1) working group on robot programming languages steps were taken to establish a common benchmark test piece for use by a-1 member states of the european community. The views of experts within Robotics Europe suggested that many classes of benchmark could be conceived, however it was considered useful to make a start by developing a physical assembly which would offer a basis for joint european experience in the comparison of robot programming systems.

Characteristics of such a benchmark were to include

(i)   compactness and portability - to fit inside a briefcase

(ii)   ability to test a variety of basic assembly operations

(iii)   universal application - must be capable of being assembled by majority of assembly robots.

The Benchmark assembly used for these tests is shown in Fig.1. The components are assembled in the assembly in the order indicated by Fig. 2 - 5. This test requires the robot to perform the basic assembly functions of:-

(i)   pick

(ii)   place

(iii)   insert

The assembly tolerances and component sizes were chosen to be within the technical specifications of the majority of assembly robots.

In order to complete this test successfully selected systems had to be able to perform several basic functions. Most also offered additional features which were useful to the assembly process.

Fig. 6 gives in a tabulated form, the relevance of these features to assembly work.

3.0   Assembly Related Robot Program Features

It is suggested that the features shown in Fig. 6 perform the following functions. They either:-

  (i)   are essential for the basic assembly process

 (ii)   improve robot assembly performance

(iii)   assist the programmer in making full use of the machine.

It is believed that basic assembly operations can be performed by a robot which has only one axis of straight line motion, control of an end effector and the ability to communicate with the world. However, such a device would lack versatility and would probably not be suitable for very many tasks.

The range of assembly operations that may be undertaken is increased if more features are introduced into the robots control system. One of the most important of these is individual step speed control which allows critical operations, such as pin insertion to be carried out at slow speeds and, hence, with a greater chance of success.

Another feature of special interest in assembly robotics is the use of tactile sensors that enable the robot's control system to be aware of the following conditions:-

  (1)   Part not present

  (2)   Excessive force required to achieve the desired position.

The conditions under which the machine took emergency actions would be preset by the robots operating system.

Other features that would influence the performance of the robot would be:-

(1)   acceleration

(2)   'settle' time

The points mentioned so far apply to the physical operating
conditions of, the robot. However, the programming system could
also be used to perform unseen tasks of benefit to the efficient
use of the machine such as:-

(1)   Diagnostics

(2)   Computational ability

(3)   Software maintenance

(4)   CAM compatability

(5)   Program backup facilities.

4.0   Comparison of Robot Languages and Programming Techniques

Not all robots have what can be termed a 'language' but this
does not mean that they cannot be programmed to perform assembly
tasks. However, the complexity of these tasks would decrease
the less comprehensive the programming system was.

It is important to stress that this paper is not concerned with
the robots themselves, only the operating systems. It was decided to cover as wide a spectrum as possible of assembly robots
ranging from the most expensive to the relatively inexpensive
machines, all of which were from the C.R.A.G. laboratory.

Those selected for comparison are:-

(1)   IBM RS2 using the AML language

(2)   Unimate PUMA using the VAL language

(3)   IBM 7535 using the AMLE language

(4)   Remek PAM

(5)   Cincinatti Milacron T^3726

(6)   ASEA IRb6

Those robots with no distinct language have instructions introduced via a key pad, menu or teach pendant.

A comparison of each robot against the assembly related robot language feature is shown in Fig. 7.

5.0 Robot Independant, Benchmark Based, Robot Language Assessment

It will be noted that the benchmark assembly is built up by a repetition of a number of basic moves. These are shown in Fig. 8. This assembly can be performed by the basic features listed as 1 - 4 in Fig. 6, providing that a facility exists for gripper interchange in order that parts of several different sizes can be accommodated. It will be noted in Fig. 9 that all of the machines tested fulfilled those criteria.

If, however, it is felt that there would be improved performance by the use of an insertion speed control facility then AMLE may not be suitable and this is shown in Fig. 10.

It may be believed that further improvements could be made by the use of a servo controlled gripper. This would save the time of gripper interchange operation. If this were to be the case, then only AML and VAL would be suitable control systems.

Finally, some operations involving close tolerance insertion may endanger the gripper or the jig if too much force is used on 'stiff' parts. To overcome this, force feedback may be necessary. If this is the case then only AML of the control systems selected could be used. This is shown in Fig. 11.

6.0 Observations

Having compared the operational characteristics of the various control systems it was felt to be necessary to compare the time required to program the robots. Obviously this is a difficult area in which to make objective judgements as different levels of programmer expertise could distort the time taken significantly. However the data given in Fig. 14 are believed to be representative.

It will be observed that AML took significantly longer to program than the other systems. This is because of the comprehensive

nature of the language and the fact that co-ordinates have to
be entered manually as there is no 'teach' button as there is
on most of the rival systems.

VAL, another comprehensive language, took nearly as long whilst
the remainder all took around 5 hours. It should be borne in
mind, however, that these systems without the facility of a servo
controlled gripper required intermediate points for gripper
changes which meant that there was not a point for point comparison but one on a task for task basis.

Another aspect that should be considered is the ease with which
a particular control system can be learnt.

To make full use of its potential, AML would require a high degree of programming expertise and a good knowledge of mathematics.
VAL, on the other hand is more user friendly and would require
less highly trained personnel.

It may well be the case that a first time user of the RS2 robot
would be advised to allow a subcontractor, such as IBM to write
the software for them.

The other robots are not taught through a language and could
easily be programmed by an operator after only a week or so of
training by the robot manufacturer.

In general the whole activity was felt to have been very reward-
ing and thought provoking. Clearly there is much more evaluation
to do on the physical benchmark itself and it is now intended to
refine the test piece and circulate the benchmark around the
member states, thereby achieving a common basis for international
cooperation in this field.

The implications of the benchmark design are numerous. Being
intentionally limited to cartezian motions in 'X', 'Y', 'Z' and
rotation only around the 'Z' axis the benchmark clearly tests
the Olivetti Sigma robot rather more comprehensively than say
the PUMA or the IBM RS2. There is therefore an implicit problem
of some importance in this direction since benchmarks requiring
more complex manipulator motions would leave a significant pro-

portion of assembly robots unable to participate in the comparison.

There can be little doubt that as assembly benchmark work progresses it will be necessary to take steps to eliminate as far as possible the limiting factors associated with diverse examples of robot hardware.

Conversely it would be valuable to see some work carried out, preferably using a suitable benchmark, which compares the performance of a robot programming language such as LM or AML when used in conjunction with a representative range of manipulators.

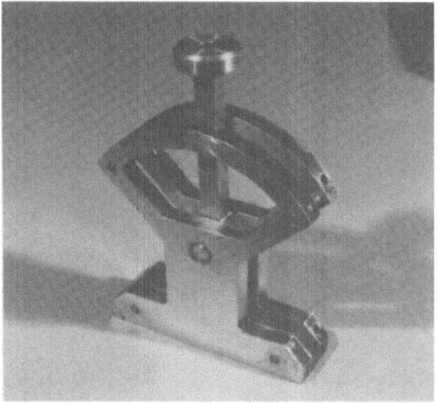

Fig. 1

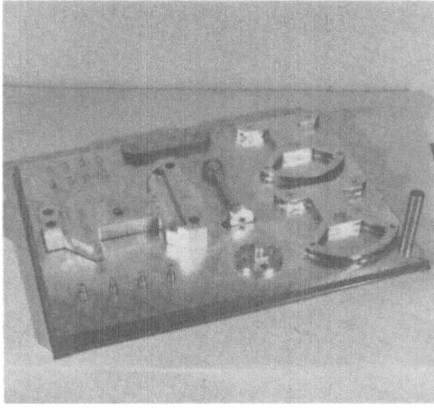

Fig. 2

Fig. 3

Fig. 4

Fig. 5

Essential	Desirable	Additional
(1) Straight line motion at least one direction	(5) Straight line motion in any direction	(13) Acceleration control
(2) Gripper control	(6) Servo control of gripper	(14) Vision facility
(3) Response to external signals	(7) Tactile sensing	(15) Software maint. checks
(4) Originate output signals	(8) Step speed control	(16) Control of 'settling' time
	(9) Editing facility	(17) CAM compatable
	(10) Computational ability	(18) Produces a listing if printer is interfaced
	(11) Program backup facility	(19) Diagnostics
	(12) Program decision making	
	(20) Direct position teach facility	

Fig. 6. Assembly related robot language and programming features

Programming Feature	AML	AMLE	VAL	PAM	T^3-726	IRb6
1	X	X	X	X	X	X
2	X	X	X	X	X	X
3	X	X	X	X	X	X
4	X	X	X	X	X	X
5	X		X		X	X
6	X		X			
7	X					
8	X	X	X	X	X	X
9	X	X	X	X	X	X
10	X	X	X		X	
11	X	X	X			
12	X	X	X	X	X	X
20		X	X	X	X	X
13	X					
14	X		X			
15	X		X			
16	X					
17	X		X			
18	X	X	X			
19	X	X	X		X	

Fig. 7

Step	Action	Comments
1	Move	General move to position above part
2	Move	Straight line move to position gripper beside part
3	Close Gripper	
4	Move	Straight line motion to withdraw part from location
5	Move	General move to position part above the appropriate place
6	Move	Straight line motion to insert part
7	Open gripper	
8	Move	Straight line motion to withdraw gripper from part

Fig. 8

		AML	AMLE	VAL	PAM	726	IRb6
Step 1	General move*	X	X	X	X	X	X
2	1, 8	X	X	X	X	X	X
3	2	X	X	X	X	X	X
4	1, 8	X	X	X	X	X	X
5	General move	X	X	X	X	X	X
6	1, 8	X	X	X	X	X	X
7	2	X	X	X	X	X	X
8	1	X	X	X	X	X	X
9	General move	X	X	X	X	X	X

Fig. 9. Basic assembly

* See fig. 6 for explanation of codes

		AML	AMLE	VAL	PAM	726	IRb6
Step 1	General move	X	X	X	X	X	X
2	1, 8	X	X	X	X	X	X
3	2, 7	X	X	X		X	X
4	1, 8, 7	X					
5	General move	X	X	X	X	X	X
6	1, 8, 7	X					
7	2	X	X	X	X	X	X
8	1	X	X	X	X	X	X
9	General move	X	X	X	X	X	X
10	End	X					

Fig. 10. Step speed control

		AML	AMLE	VAL	PAM	726	IRb6
Step 1	General move	X	X	X	X	X	X
2	1, 8	X	X	X		X	X
3	6, 7	X					
4	1, 8, 7	X					
5	General move	X	X	X	X	X	X
6	1, 8, 7	X					
7	6	X		X			
8	1	X	X	X	X	X	X
9	General move	X	X	X	X	X	X
10	End	X					

Fig. 11. Force sensing

Control System	AML	VAL	AMLE	PAM	T^3-726	ASEA IRb6
Time Taken (hrs)	10	8	7	5	5	5

Fig. 12

References

1. Internal Paper on Benchmark Design. Robotics Europe -
   Working Group on Robot Programming. December 1983.

2. Gruver, W.A. et al: Evaluation of Commercially Available
   Robot Programming Languages. Proc. 13th International
   Symposium on Industrial Robots. April 1983. Robotics Int.
   if SME. Dearborn Michigan.

3. Bonner, S. and Shin, K.G.: A Comparative Study of Robot
   Languages. Computer December 1982.

4. CAM-I Robotics Software Project 1983. PR-82-ASPP-01.1
   CAM-I INC. Arlington, Texas, U.S.A.

M. Andreasen, S. Kähler, T. Lund

Design for Assembly

1983. 189 pages.
Cooperation with IFS (Publications) Ltd., U.K.
ISBN 3-540-12544-2

Decade of Robotics

Special Tenth Anniversary issue of
The Industrial Robot magazine
Editor-in-Chief: **J. Mortimer**
Editor: **B. Rooks**
1983. Numerous figures. 168 pages. Cooperation with IFS
(Publications) Ltd., U.K. ISBN 3-540-12545-0

P. Dransfield

Hydraulic Control Systems – Design and Analysis of Their Dynamics

1981. VII, 227 pages. (Lecture Notes in Control and Information Sciences, Volume 33). ISBN 3-540-10890-4

Flexible Automation in Japan

Editor: **J. Hartley**
1984. VI, 264 pages. IFS (Publications) Ltd., U.K.
ISBN 3-540-13499-9

A. Gomersall

Machine Intelligence

An International Bibliography with Abstracts of Sensors in
Automated Manufacturing
1984. 122 figures. VII, 232 pages. IFS (Publications) Ltd.,
U.K. ISBN 3-540-13191-4

J. Hollingum

Machine Vision

The Eyes of Automation
1984. 20 figures. Approx. 100 pages. IFS (Publications) Ltd.,
U.K. ISBN 3-540-13837-4

Ingersoll Engineers

The FMS Report

Editor: **J. Mortimer**
2nd revised printing. 1984. 184 pages. IFS (Publications) Ltd.,
U.K. ISBN 3-540-13556-1

Springer-Verlag
Berlin
Heidelberg
New York
Tokyo

Mechanical Hands Illustrated

Compiled by **I. Kato**
Editor of English Version: K. Sadamoto
1982. XXI, 214 pages.
Tokyo: Survey Japan
ISBN 3-540-13060-8

C. Morgan

Robots
Planning and Implementation

1984. XI, 195 pages.
IFS (Publications) Ltd., U.K.
ISBN 3-540-12584-1

T. Müller

Automated Guided Vehicles

1983. V, 290 pages.
IFS (Publications) Ltd., U.K.
ISBN 3-540-12629-5

Programmable Assembly

Editor: **W. B. Heginbotham**
1984. XIII, 349 pages. (International Trends
in Manufacturing Technology)
IFS (Publications) Ltd., U.K.
ISBN 3-540-13479-4

Robot Vision

Editor: **A. Pugh**
1983. XI, 356 pages. (International Trends in
Manufacturing Technology)
IFS (Publications) Ltd., U.K.
ISBN 3-540-12073-4

Springer-Verlag
Berlin
Heidelberg
New York
Tokyo

Scientific Fundamentals
of Robotics

Volume 1
M. Vukobratovic, V. Potkonjak
Dynamics of Manipulation Robots
Theory and Application
1982. 149 figures. XIII, 303 pages. (Commu-
nications and Control Engineering Series)
ISBN 3-540-11628-1

Volume 2
M. Vukobratović, D. Stokić
Control of Manipulation Robots
Theory and Application
1982. 111 figures, XIII, 363 pages. (Commu-
nications and Control Engineering Series)
ISBN 3-540-11629-X

Volume 3
M. Vukobratović, M. Kirćanski
Kinematics and Trajectories
Planning of Manipulation Robots
1984. Approx. 190 pages. (Communications
and Control Engineering Series)
ISBN 3-540-13071-3

Volume 4
M. Vukobratović, N. Kirćanski
Real-Time Dynamics of
Manipulation Robots
1984. Approx. 280 pages. (Communications
and Control Engineering Series)
ISBN 3-540-13072-1

Volume 5
M. Vukobratović, D. Stokić, N. Kirćanski
Non-Adaptive and Adaptive Control
1984. Approx. 330 pages. (Communications
and Control Engineering Series)
ISBN 3-540-13073-X

Volume 6
M. Vukobratović, V. Potkonjak, I. Nikolić
Computer Aided Manipulation
Robots Design
1984. Approx. 330 pages. (Communications
and Control Engineering Series)
ISBN 3-540-13074-8